PRACTICAL LABORATORY INFORMATION MANAGEMENT FOR SCIENTISTS & ENGINEERS

Prentice Hall Laboratory Lotus® Series

Mezei *Laboratory Lotus®: A Complete Guide to Instrument Interfacing*
Mezei *Practical Spreadsheet Statistics and Curve Fitting For Scientists and Engineers*
Mezei *Practical Laboratory Information Management For Scientists and Engineers*

PRACTICAL LABORATORY INFORMATION MANAGEMENT FOR SCIENTISTS & ENGINEERS

Louis M. Mezei

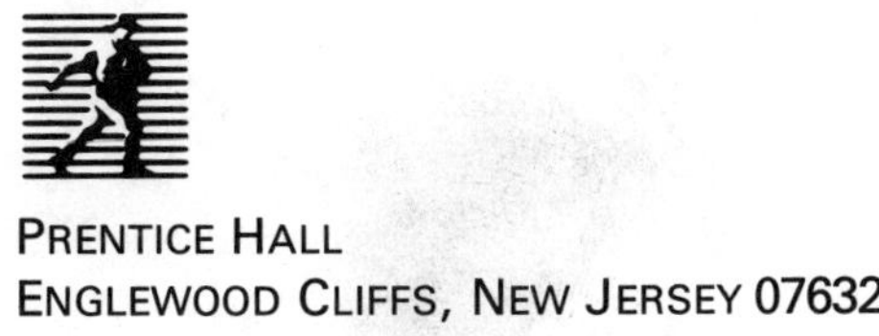

Prentice Hall
Englewood Cliffs, New Jersey 07632

Library of Congress Cataloging-in-Publication Data

Mezei, Louis M.
Practical laboratory information management for scientists & engineers / Louis M. Mezei.

p. cm. — (Prentice Hall laboratory Lotus series)
Includes bibliographical references and index.
ISBN 0-13-678186-1
1. Physical laboratories—Management—Data processing.
2. Information storage and retrieval systems—Science.
3. Information storage and retrieval systems—Engineering.
4. Information resources management. 5. Lotus 1-2-3 (Computer program) 6. Electronic spreadsheets. I. Title. II. Series.
Q180.57.M49 1992 91-13882
502'.85'5369—dc20 CIP

Editorial production supervision: Barbara Marttine
Cover design: Karen Stephens
Prepress buyer: Kelly Behr
Manufacturing buyer: Susan Brunke
Acquisitions editor: Michael Hays

The publisher offers discounts on this book when ordered in bulk quantities. For more information, write:

Special Sales / Professional Marketing
Prentice-Hall, Inc.
Professional & Technical Reference Division
Englewood Cliffs, New Jersey 07632

Printed in the United States of America

10 9 8 7 6 5 4 3 2 1

ISBN 0-13-678186-1

Prentice-Hall International (UK) Limited, *London*
Prentice-Hall of Australia Pty. Limited, *Sydney*
Prentice-Hall Canada Inc., *Toronto*
Prentice-Hall Hispanoamericana, S.A., *Mexico*
Prentice-Hall of India Private Limited, *New Delhi*
Prentice-Hall of Japan, Inc., *Tokyo*
Simon & Schuster Asia Pte. Ltd., *Singapore*
Editora Prentice-Hall do Brasil, Ltda., *Rio de Janeiro*

It's easy to keep track of a couple of anything. But when you have millions, then you need some help.--Lou

Contents

Preface

How I Got Hooked On the Lotus® Spreadsheets and Programming Language

Have you ever noticed how scientific software manufacturers corral you into a one-inch square box and then charge you megabucks for the box?

Have you ever noticed how commercial software manufacturers try to make you do it their way, or not at all?

Are you having a hard time finding programs that reduce and calculate your data the way that ***YOU*** want them evaluated?

Are you tired of buying new software, only to be directed through yet another mountain of manuals every time you want to do something a little different?

Do you get confused trying to remember which commands go with which piece of software; leaving you with a strong desire to try to find one system that would consolidate your efforts into a single way of doing things?

Are you disappointed with the cost/benefit ratio of Laboratory Information Management Systems (LIMS)?

Trying to find an amicable, all-encompassing answer to the above problems prompted me to look into spreadsheet programs. At first, I was awed at the number of things that I thought I would need to learn. What's worse, all of the books written on spreadsheets were written from a business standpoint. After buying about 16 books on Lotus® spreadsheets and trying to weed through such foreign topics as "rate of return," "sales forecasts," "commission schedules," and the like, I became dismayed. But, I saw a glimmer of hope. Lotus 1-2-3® and Symphony® seemed to have all of the features that

I needed to reduce and graph my data. All that I needed to do was to transform the information in my books from a business focus to a scientific standpoint and try to find a way to use the features in a way that would address the kinds of programming and spreadsheet activities that a scientist could benefit from.

My incentives were numerous. I was buying expensive software that was designed by people who had never worked in a laboratory environment. The software was never a good "fit" for how things were really done in laboratories. I had doubts that software designers had ever caught on to the fact that every experiment performed in a laboratory setting was unique. This uniqueness is especially true in research laboratories. In addition, data reduction tends to be highly variable. In fact, most researchers look at their data from many different angles...the number of angles varying with the inverse of the data's quality.

To make a long story short, I have found Lotus spreadsheets to be a cost effective way to beat the problems associated with purchased software. I was able to find a method to talk to instruments directly from a spreadsheet, most often ***WITHOUT the use of Lotus Measure®***. Once the data were in the spreadsheet, Lotus solved all of my data reduction and graphics problems. Lotus Symphony and an add-in called "@BASE" solved all my record-keeping problems.

Lotus is a very rich environment that can solve your data acquisition, reduction, graphic and record-keeping problems. In fact, the Lotus environment, programmed correctly, can provide an average-sized laboratory with a system that compares favorably with expensive ($10,000 to $100,000) Laboratory Information Management (LIM) systems.

You do not need to learn the hundreds of Lotus program commands and special-purpose formulae to use it effectively. You also do not need to spend a lot of time learning how to connect an instrument to a personal computer. My first book, ***Laboratory Lotus®: A Complete Guide To Instrument Interfacing*** teaches you quick and easy ways to get data into your spreadsheet from your instruments. My second book, ***Prentice Hall Laboratory Lotus® Series™: Practical Spreadsheet Statistics and Curve Fitting for Scientists and Engineers***, teaches you efficient and flexible ways to quickly perform data reduction on data, whether the data are from instruments or not. This book, ***Prentice Hall Laboratory Lotus® Series™: Practical Laboratory Information Management for Scientists and Engineers***, teaches you how to store the final answers in an accessible form that will allow you to retrieve the exact information you need...quickly and conveniently.

From robots to doughnuts (and everything in between), if your central mission is the collection and reduction of data, the ***Prentice Hall Laboratory Lotus® Series*** can help you. Armed with the first book, you can control just about everything from a robot that automates the most sophisticated genetic engineering project to a texture analyzer that measures the adhesion, bloom strength, crispness, and tackiness of your doughnuts to make sure that they are done and have not become stale. And that information can be very important at coffee-break time.

The second book will help you make your data come alive with sophisticated statistical and curve fitting data reduction, even if you have just typed the data into the spreadsheet. When your reports contain statistical characterizations, distribution analyses, matrix mathematics, linear and curvilinear regressions, nonlinearly fitted curves, spline interpolations, ANOVA tables, nonparametric statistics, etc., you can issue reports and draw conclusions with increased confidence.

This book will help you store the data in readily retrievable form. After reading this book and learning some programming techniques, you will be able to write ***your own*** software tailored specifically to ***your*** data storage needs in just a couple of hours.

LOUIS M. MEZEI

1

Introduction

My first two books in this series taught you how to merge scientific instruments, your personal computer, and either a Lotus® 1-2-3® or Lotus Symphony® spreadsheet to produce a single, powerful unit that met your exact automation, data reduction, and graphics needs. This book will show you how to store and retrieve the reduced data.

This chapter will provide some preliminary definitions of terms, an overview of the book, and the software and equipment that you will need.

SOME USEFUL DEFINITIONS

When most people think of Lotus 1-2-3 and Symphony, they think of an electronic spreadsheet. Spreadsheets are excellent tools to capture data from instruments and reduce the data to a form that can be used to make important decisions.

In most laboratories, the storage of experimental results in a form that can be readily retrieved by a wide variety of individuals at later dates is key to the success of focused research, development, and quality control.

The organized storage of information for later retrieval is called ***database management***. A database management system provides many useful functions that aid the review of data. For example, a database management system performs functions such as the sorting, selecting, searching, revising and printing of data. This section will provide definitions for a few of the basic database terms that will be used throughout this book. These terms will be defined in more detail in later chapters, but cursory explanations will provide fundamental descriptions to make understanding easier prior to rigorous treatment.

Records

A ***record*** is simply a collection of related information. You probably use records all the time without thinking of them as records. A Rolodex card is a good example of a record.

If you understand how a Rolodex card relates to a computerized record, you can more easily understand the concept of a computerized record. Like a Rolodex card, a computerized record consists of structured information. A Rolodex card contains information about one person or thing. Each item of information on a card is called a ***field***. On a Rolodex card, the fields may be listed as follows:

MeziLu Productions
Louis M. Mezei
40815 Ondina Ct.
Fremont, CA 94539
(415) 555-1234

COMPANY
NAME
ADDRESS
CITY
STATE
ZIP CODE
PHONE NUMBER

In the laboratory setting, a Rolodex card may contain fields that hold information about a single sample that was submitted for testing. For example, a record may contain fields for the sample's barcode number, type of assay, the name of the requester, department code numbers, storage location, results, status, reagents used during assay, etc.

The figure in this section shows a typical Rolodex card. In database terminology, every record (card) contains the same fields (items of information) and has the same structure (organization). Together, these records make up the file, or ***database***.

Databases

A ***database*** is a collection of related records that is neatly organized so that it can be accessed and manipulated in a variety of ways. In effect, a database is a collection of many Rolodex cards. Another common manual database example is a file cabinet containing file folders.

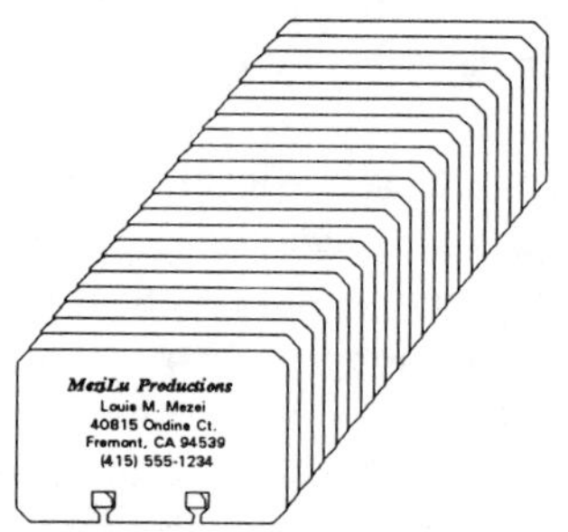

Computerized databases are similar in concept to manual databases. However, computer databases are essentially ***tables*** of logically related information stored in a file on disk. Information stored in computerized databases is highly structured and follows strict rules. This tabular approach makes for some very important improvements on manual databases in terms of ease of data manipulation and retrieval.

Index Files and Keys

Each time a new record is added to a database, it is appended to the bottom of a file. If a record is to be found, the file must be searched from the beginning of the file until the record is found. This process is time-consuming, especially for large files. If the

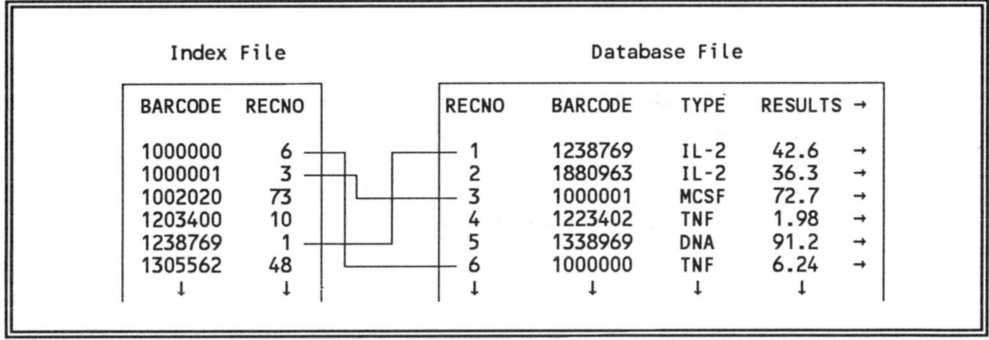

Index File

BARCODE	RECNO
1000000	6
1000001	3
1002020	73
1203400	10
1238769	1
1305562	48
↓	↓

Database File

RECNO	BARCODE	TYPE	RESULTS →
1	1238769	IL-2	42.6 →
2	1880963	IL-2	36.3 →
3	1000001	MCSF	72.7 →
4	1223402	TNF	1.98 →
5	1338969	DNA	91.2 →
6	1000000	TNF	6.24 →
↓	↓	↓	↓

Figure 1-1
Index File

database is to be ordered in a particular way, it must be sorted each time a new record is added. This process is also time-consuming.

Index files can dramatically decrease sorting and searching times. This efficiency gives you substantially faster access to database records.

Database index files are special files that contain information about the location of records in a database. A database index is like the index for a book. A book index lets you identify the page where a topic is discussed without searching through the entire book cover to cover. A database index uses record numbers instead of page numbers, but the concept is the same. Using an index file and a ***key*** value, you have quicker access to records in the database. An index ***key*** is a field in an index file that points to the locations of records in a database file. The index key determines the order that the records in a data file are displayed. The index key is also used to search for records that need to be displayed or updated.

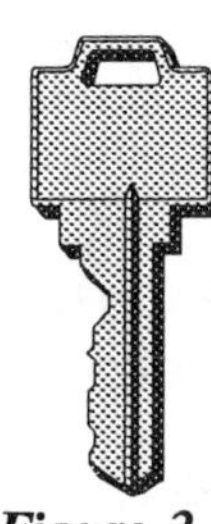

Figure 3

Adding a new record to a database with an index file involves two files. The new record is added to the end of the database as usual. In addition, the key value field and a pointer to the database record are added to the index file which is maintained in sorted order. Because two files are being updated rather than one, adding a new record to the indexed database takes slightly more time than updating a database without an index file. However once an index file has been created, the database can be accessed very quickly because the smaller, ordered key file provides a direct pointer to the record in the larger, unordered data file.

Figure 1-1 illustrates how an index file works. The index file contains a single pair of entries for each record in the database: a key value (e.g., a barcode) and a record number. Having a limited amount of compact data in a file makes the file much smaller in size and substantially improves the time it takes to search the file for data. Maintaining the file in sorted order also improves search time. For these two reasons, searching through index files is much faster than searching through large, unordered database files containing immense amounts of data.

In Figure 1-1, the complete record for any sample in the database file can be quickly located by

- first searching the index file to find the sample's BARCODE number,
- obtaining the sample's record number from the index file,
- immediately skipping to the offset in the database file corresponding the record number, and
- retrieving the pertinent information.

You can have more than one index per database file. It is often useful to create several index files for a single database. You can also have ***compound indexes***, where the index key consists of more than one field. For example, in addition to the BARCODE index, a second index file containing a different set of keys (e.g., STATUS and REQ_DT), might be helpful to speed preparation of end-of-month reports. If more than one index is open for a database, the ***primary index*** is the one containing the key(s) used to find records. The first index opened for a particular database is the ***primary index*** until another index is selected as the primary index. The primary index is important because it determines which records are retrieved from the database when a search is performed and the order in which records are displayed.

For the applications in this book, once you have created index files for databases, the index files are automatically kept up-to-date and arranged in ascending alphabetical or numerical order. That is, when records are modified, added, or deleted, index files are amended to maintain the proper sort order. This arrangement provides maximal efficiency for data retrieval.

Relational Databases

Strictly speaking, a ***relational database*** is a database that is structured so that fields of information are logically related to unique records. Each field describes a particular attribute of the record. For example, unique records might be specified by sample barcode numbers. A relational database describing sample attributes would be a two-dimensional table (often called a "flat file"), each of the table's columns being a field, and each field describing a particular attribute of the unique barcode number. In a relational database, data are accessed on the basis of their ***relationship*** to other data. Some example fields in a relational database might be

BARCODE NUMBER
DATE OF REQUEST
NAME OF REQUESTER
TYPE OF ASSAY
RESULTS
REAGENTS USED
STATUS

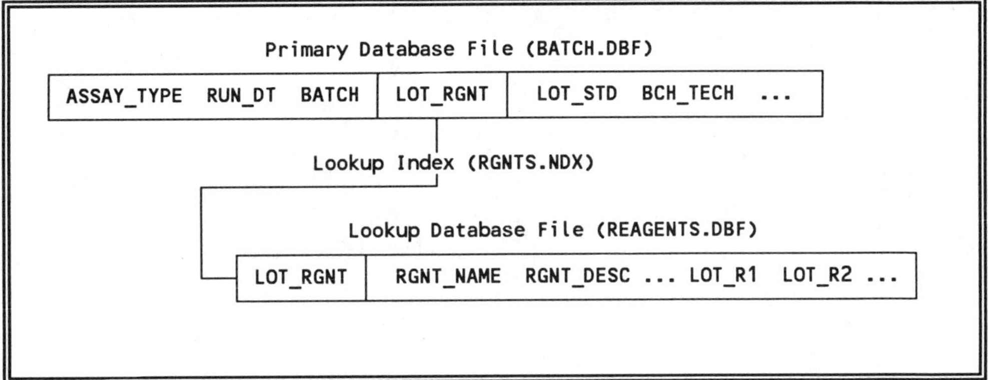

Figure 1-2

However, this rigorous definition of a relational database has limited usefulness to scientists and engineers. Therefore, for the purposes of this book, a broader (and more conventional) definition of a relational database will be used. For this book, an ***extended relational database*** is a group of two or more related data ***files*** that work together in a special way. As illustrated in Figure 1-2, information that might otherwise be stored in a single large file with many fields is instead organized in several files, each with fewer fields. The files are ***joined*** together (***linked***) so that information can be shared among the files. In a sense, one data file can look up information in one or more other data files. When individual database files are linked together to extend the ready availability of related information, the resulting system appears to be a single, massive file, containing many fields. This system is often referred to as a ***virtual database***.

A virtual database consists of a primary file and one or more lookup files. The information that is unique to each record is stored in a file, called the ***primary file***. Information that is shared by many records is stored as a single record in a separate file, called the ***lookup file***.

To join files, at least one field must be common to both files being joined (e.g., LOT_RGNT in Figure 1-2). Further, when databases are joined, the common field must have the ***exact*** same name in both the primary and lookup files. The ***lookup*** file must be indexed on the field that is common to both files. The databases are joined, using the lookup index file as an active link between the two database files. The index file link provides the extended lookup capabilities.

With an extended relational database, information is entered once, but can be shared among several records. The ability to look up information eliminates unnecessary redundancy of data, saving time and disk space. For example, in a relational database used for sample processing, information about the reagents used can be stored in a separate file. If a set of reagents are used for many assay runs, information such as date of preparation, shelf life, preparer, etc., can be brought in from a reagent (lookup) file rather than repeated in the record for each assay run. The reagent file is "joined" to the

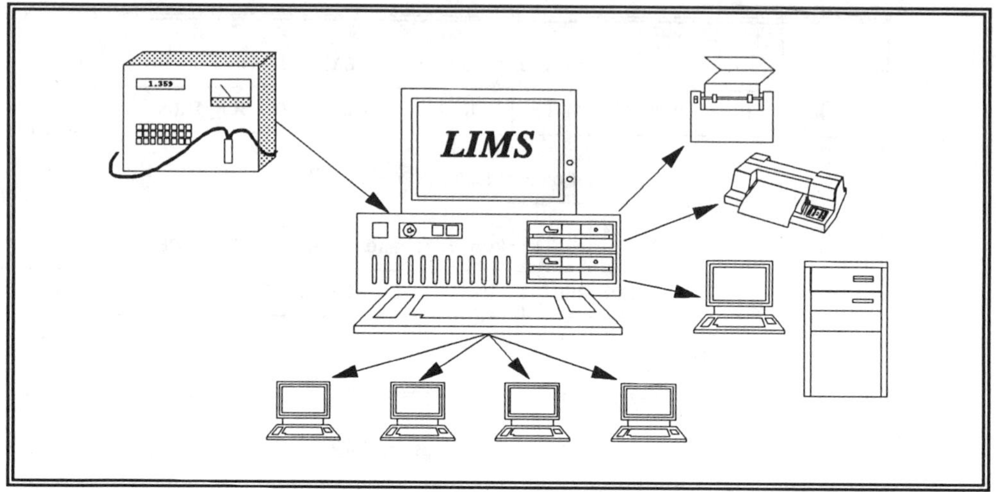

Figure 1-3
A Typical LIM System

run identification (primary) file. Likewise, if several runs have a common standard curve, a second lookup file could be joined to the assay run identification file to supply information on the standard curve results.

Laboratory Information Management Systems

A ***DataBase Management System*** (DBMS) is a software program for managing the creation of, storage for, updating of, and access to data stored in a database.

A ***Laboratory Information Management (LIM)*** system is a database specifically tailored to an individual analytical laboratory's needs. The main purpose of a LIM system is to manage the flow of information in a laboratory and thereby provide high quality results, fast turnaround, optimal resource allocation, and cost effectiveness.

Thus, LIM systems integrate sample information with results produced from analytical techniques and processes to maintain data integrity and increase the speed of reporting results. The LIM system database controls the flow of information. A sample is logged into the system as it enters the laboratory, tests are performed, data are obtained, results are calculated and entered into the database, and reports are issued. By integrating sample information (e.g., source, batch number, sampling information) with the results from instruments and automatically issuing reports, the efficiency and accuracy of a laboratory can be increased because

- the tedium of manually filing test results can be eliminated.
- sample reports can be easily and quickly obtained.

- important data can be highlighted.
- trends can be graphically exposed.
- fast access to historical records can be obtained.
- more time can be spent on interpretation of data.
- administrative tasks can be drastically reduced.

In many respects the name Laboratory Information Management System is a misnomer applied to many contemporary LIM systems because they do not manage information, they manage data. It is a goal of this book to help you create your own custom LIM system that will manage both data and information.

HOW THIS BOOK IS ORGANIZED

This book is progressive in format, but contains a certain degree of redundancy (when necessary). The remainder of this chapter is devoted to stating any assumptions I've made and provides you with all the information you need to complete the objectives of the remaining chapters successfully.

This book is roughly divided into three major sections:

- The first section (Chapters 2 and 3) deals with design considerations for information management systems, databases, and spreadsheet organization.
- The second section (Chapters 4 and 5) deals with constructing databases, creating index files, joining databases to provide relational capabilities, and modifying existing databases.
- The third section (Chapters 6 through 10) deals with the programming that you will need to know to merge Lotus spreadsheets, database tables and disk management systems to form a LIM system tailored to your needs.

The following is a brief overview of each chapter:

- Chapter 2 explains how to design an extended relational database system for your laboratory. It also provides a description of the example database system that will be used throughout the remainder of the book.
- Chapter 3 explains how to plan and organize your data for efficient and reliable manipulation. To this end, Chapter 3 presents the concept of "spreadsheet templates" and how templates can be used as pipelines to facilitate data flow between data acquisition, data reduction, databases, reports, graphics, etc. Creation of a user interface and report forms is also explained.
- Chapter 4 explains how to configure @BASE and the @BASE Option Pac (add-in application programs that manage databases that reside on disk). It also explains how to create databases and database definitions.
- Chapter 5 explains how to use the @BASE Option Pac to create links between files (to provide relational database capabilities), index files, and computed fields.
- Chapter 6 will teach you the programming skills you will need to know to create

an integrated information management system. A comparison to BASIC language is made for those who know BASIC. However, the focus is on Lotus programming commands and techniques. If you do not have a copy of ***Laboratory Lotus®: A Complete Guide To Instrument Interfacing***, then it is important for you to read and understand the information provided in Chapter 6 before proceeding to chapters which follow.

- Chapter 7 describes an easy method that you can use to plan program flow for your LIM system. This method is based on creating a hierarchy of interconnected menus instead of creating flow diagrams. Chapter 7 also explains how to create macro programs that automatically execute at system start-up, determine the release of Lotus software being used, configure the system, and launch the menu system.
- Chapter 8 explains how to create re-usable software tools for the LIM system. By creating multiple purpose software subroutines, your programs will consume less memory, execute more efficiently, and provide more reliable results.
- Chapter 9 provides example programs that show how to use the generic subroutine utilities described in Chapter 8 to move data between database files and spreadsheet templates.
- Chapter 10 is a summary chapter that provides pointers on designing databases, building templates, programming, troubleshooting problems, etc.
- Appendices A, B, and C present concise explanations of the Lotus operators, special purpose functions, and program commands used throughout the book.
- Appendix D presents concise explanations of @BASE @db and @ndx @functions. Appendix E presents explanations of @BASE operating functions.

All of the chapters in this book point out ways to avoid some of the common pitfalls found in Lotus. They also show you how to avoid problems arising from quirks in Lotus' operation, and what to do if you fall into a trap.

If you are an experienced Lotus user, this book will show you how to push Lotus to its limits by taking worn out, old Lotus tools that have been collecting dust and use them in innovative ways to perform some very clever tasks.

ASSUMPTIONS ABOUT YOUR ABILITIES

I wrote this book for the person who has access to a copy of Lotus 1-2-3 and/or Symphony and has at least a beginner's knowledge of how to use it. A fundamental discussion of Lotus spreadsheets and their components can be found in subsequent sections of this chapter. A more thorough discussion of these topics can be found in one of the references listed at the end of this chapter.

I also assume that you have a real desire to learn how to perform some basic programming techniques to solve your data handling challenges.

Finally, I do not intend for this book to be a comprehensive text on Laboratory Information Management (LIM) systems. This book is a **practical** guide on **how to**

create your own LIM system for real-life data.

ASSUMPTIONS ABOUT YOUR SOFTWARE AND PERSONAL COMPUTER

Spreadsheet Software

I developed and tested all of the examples in this book using Lotus 1-2-3 (Releases 2.0, 2.01, and 2.2) and Symphony (Releases 1.2, 2.0, and 2.2). Contrary to previous books in this series, the add-in application software required to implement the features of this book do not work with Symphony, Release 1.1. Therefore, if you are using Release 1.1, you will need to upgrade to a newer version.

If you are in the process of deciding whether to purchase or use Lotus 1-2-3 or Symphony for your system, consider the following: This book contains programs that supplant the Symphony database environment. Previous books in this series consumed relatively small amounts of memory. However, large-scale database management is different in terms of memory usage. For example,

- database spreadsheets tend to be large.
- database macro programs tend to be large.
- several, large, add-in programs are needed.
- index files are used.

All of these categories are stored in conventional memory. This storage can consume conventional memory very quickly. For this reason, Lotus 1-2-3 has an advantage because it consumes much less memory than Symphony. Another advantage of Lotus 1-2-3 is that it is much faster than Symphony. Because the database environment of Symphony will not be used (and, in fact, will be disabled), I would recommend Lotus 1-2-3 for this application.

With Lotus 1-2-3, Release 3, Lotus Development made some substantial changes in the way that add-in application software programs interface with the spreadsheet program. For this reason, third-party developers have had to make significant changes to their add-in software and re-issue it, thus causing a delay in the availability of add-in software for Release 3. ***The next section describes add-in software that is required for this book. At the time of this book's publication, the add-in software necessary to implement the concepts of this book was unavailable for Release 3.*** In anticipation of the impending availability of, and with the faith that the software will function identically to, corresponding add-ins for prior releases of Lotus 1-2-3, Release 3 will be treated as if it were indistinguishable from prior releases.

Because the Lotus macro commands and @functions are fully established, there should not be any changes in how they are invoked or how they perform (other than additions required to support the multiple layering of spreadsheets). For this reason, the information and examples given in this book should be generally applicable to all future releases of the two Lotus programs.

Add-in Software

Lotus 1-2-3 and Symphony have some database management capabilities built in. However, these capabilities are too limited to be used as a basis for an effective laboratory database management system. For example,

- The number of records that can be handled is limited by the number of rows in the spreadsheet and available memory. For Lotus 1-2-3 and Symphony, the maximum number of records that you can conveniently handle is 8192.
- The speed of operation is unacceptably slow, especially as the database grows.
- Dissemination of data to other terminals in a network is awkward.

These limitations make the Lotus programs unacceptable when it comes to using them alone to control the database of a LIM system. For this reason, you will need to add the ability to dynamically store and access records on disk. This additional functionality could conceivably be accomplished using the existing Lotus programming language. However, creating an efficient and reliable system from these commands is both difficult and time-consuming.

For these reasons, the foundation for the applications described thoughout this book is a database management system that can be purchased from Personics Corporation (63 Great Road, Maynard, MA 01754, (800) 445-3311) or directly from the author using the order form at the end of this book. The products that you will need to purchase are called @BASE and @BASE Option Pac. These programs have the power of stand-alone database programs like Ashton Tate's® dBASE III®, but operate inside Lotus 1-2-3 and Symphony.

With @BASE, data are stored on disk, using exactly the same file format as dBASE III and dBASE III plus, thus providing an almost unlimited capacity to the number of records that you can store (up to 32 MB, depending on hard-disk space). @BASE also provides very fast access to the information. In fact, the access time is up to 3.7 times faster than dBASE III, which in turn is orders of magnitude faster than the Lotus programs.

The @BASE Option Pac provides a joined-relational database and indexing capabilities.

With the bi-directional data transfer provided by @BASE (from spreadsheet to database and from database to spreadsheet), you get the best of both worlds. Your spreadsheet gives you tools for data acquisition and analysis, while @BASE gives you enormous data storage capacity.

@BASE and @BASE Option Pac also provide some other benefits to your Lotus-based LIM system that Lotus alone does not readily provide. They are:

- Concurrent accessing of multiple data files.
- Powerful Query/Selection commands.
- Database packing (deleting marked records).

- Creating and maintaining index files that are fully compatable with the dBASE NDX format.
- Quickly sorting a database without physically rearranging the records on disk.
- Joining (linking) two or more database files to create a single "virtual" database with relational properties.
- Creating "computed" fields whose values are derived from other fields in the database.

If you have Lotus 1-2-3 (Release 2.0, 2.01, or 2.2) or Symphony (Release 1.2, or 2.0), you will need to purchase both @BASE and the @BASE Option Pac.

If you have Symphony, Release 2.2, you already have @BASE built in. However, you will need to purchase the @BASE Option Pac. When ordering, make sure that you specify the release of Symphony that you have.

DOS Version

LIM systems typically have multiple index and database files concurrently open. For this reason, you will need a version of DOS that can readily handle multiple open files. This capability necessitates the use of DOS version 3.2 or above. After DOS version 3.2, the number of open files can be specified using a FILES command in CONFIG.SYS files. The maximum number of open files is 255. However, there is a limitation that you need to be aware of. Although DOS allows up to 255 files to be open, only 20 files can be accessed by each application. Thus, Lotus and any add-in application programs together cannot access more than 20 files, regardless of how many files are specified in a FILES command of a CONFIG.SYS file. This limitation will affect the way that you design your system and is addressed further in Chapter 2.

To verify that minimum requirements have been met, check your CONFIG.SYS file to ensure that a FILES command is present and set to at least 20 open files. If not, use your word processor to change the FILES statement and re-boot your computer.

Personal Computer Equipment

The computers that I used were all IBM®-based. I have a Personal System/2® Model 80, an AT®, and an XT™. All of the programs worked on all three computers. The XT proved a little slow in a couple of instances, so I used plug-in accelerator boards to boost its speed.

The first board that I used was an Orchid Technology PCturbo 286e plug-in accelerator board. (The PCturbo 286e is available from Orchid Technology, 45365 Northport Loop West, Fremont, CA, 94538.) The speed increase provided by the PCturbo 286e was dramatic. The PCturbo 286e model that I own is an 8 MHz, zero wait-state 80286 CPU with cache memory. Because the board also uses the XT's existing 8088 microprocessor to handle input and output, the board runs rapidly through calculations and video displays. One drawback to the PCturbo 286e is that it is not compatible with EGA monitors.

The second board I used was an Intel® Inboard™ 386/PC in the XT. (The Inboard 386/PC is available from Intel Corporation, 5200 NE Elam Young Parkway, Hillsboro, OR 97124-9987.) This board has a 80386 microprocessor working at 16 MHz and provides all of the advanced features of the 80386 microprocessor. This capability means that the Intel Inboard is fully compatible with Lotus 1-2-3, Release 3, and that the board can perform calculations up to 10 times faster than the XT alone. Additionally, you can have up to 4 megabytes of high performance 32-bit extended memory for efficiency when using very large spreadsheets. More importantly, the Inboard is fully compatible with EGA monitors.

Drawing from my experience, you should be able to use any IBM personal computer or close compatible that supports Lotus. Use one with a speed at least as fast as an eight megahertz AT. The difference in price between an AT and XT is inconsequential, and the AT's speed will make calculations much faster. If you already have an XT or compatible, you may want to consider a coprocessor board for some applications.

Hard Drive. A hard drive is a necessity. Do not try to use either Lotus program without a hard drive. The current cost of hard drives will make it one of your best investments. The difference in price between a 10- or 20-megabyte hard drive and a 30 or 70-megabyte hard drive is so insignificant that I recommend you buy ***at least*** a 30 as you will soon surpass 20 megabytes.

RS-232 Communications Port. If you will also be interfacing instruments to your computer, the last necessity is at least one serial RS-232 communications port. If you have two, however, two serial ports can automate two instruments. And, if you will be buying a serial printer, the second serial port is an absolute must for interfacing instruments because the printer will occupy the first serial port.

Graphics Equipment. If you plan on using graphics, make sure that your monitor and printer can handle graphics. With new releases of Lotus products, an EGA or VGA

	1-2-3 2.0/2.01/2.2	SYMPHONY 1.1/1.2	SYMPHONY 2.0/2.2
INTEGERS	CONV	CONV	CONV
DECIMAL NUMBERS	EMS	EMS	EMS
LABELS (TEXT IN CELLS)	EMS	EMS	EMS
MACRO PROGRAMS	EMS	EMS	EMS
FORMULAE	EMS	EMS	EMS
@FUNCTIONS	EMS	EMS	EMS
RANGE NAMES	CONV	CONV	EMS
BLANK (ERASED) CELLS	CONV	CONV	CONV
ADD-IN APPLICATION PROGRAMS	CONV	CONV	CONV
SETTINGS SHEETS	CONV	CONV	EMS
LOTUS 1-2-3/SYMPHONY	CONV	CONV	CONV
PRINT SETTINGS	CONV	CONV	EMS
LOTUS TEXT (MESSAGES)	CONV	CONV	EMS
MEMORY RESIDENT PROGRAMS	CONV	CONV	CONV

Table 1-1

monitor is well worth the extra cost.

Math Coprocessor. Math coprocessors are specialized microprocessors that perform floating-point calculations. Without a math coprocessor, your computer performs floating-point calculations using integer mathematics and requires many calculations to produce a single floating-point result. If a math coprocessor is present, floating-point calculations can be executed with a single instruction, thus improving the speed with which your spreadsheet recalculates and programs perform. This speed increase is usually between 10% to 50%, but depends on the application. Scientific data reduction applications typically involve large macro programs and/or trigonometric functions. These two categories are the most common to benefit from the addition of a math coprocessor.

Expanded Memory. In previous books of this series, I recommended that if you were going to be dealing with a very large amount of data, that you might want to consider adding expanded memory to your computer. I cautioned that expanded memory was slower than conventional memory and because of the slower speed, expanded memory should only be used for very large spreadsheets.

This book, on the other hand, requires add-in application programs that utilize substantial amounts of conventional memory. This consumption of memory can substantially decrease the amount of data that you can handle in your spreadsheets. This subsection will provide you with some background on memory utilization to help you make your choice. Later in this chapter, I will provide some tips on actions you can take to make more conventional memory available and perhaps eliminate the need for expanded memory.

Lotus®, Intel®, and Microsoft® Corporations have jointly developed a specification that allows programs to use memory above the standard 640K maximum. Memory boards conforming to this specification are referred to as *E*xpanded *M*emory *S*pecification (EMS) boards. Lotus 1-2-3, Releases 2 and 3, and Symphony can take advantage of the memory on expanded memory boards. These boards are marketed by Intel, Quadram®, AST®, and many others. You can use up to four megabytes of EMS memory from a Lotus spreadsheet.

If expanded memory is present in your computer, Lotus will use the EMS memory to store formulae, floating point (decimal) numbers, and labels. Integers are stored in standard memory. Table 1-1 outlines the type of memory (CONVentional or EMS) where various types of information reside. As you can see, there are subtle, but very important, differences in how the various Lotus programs utilize expanded memory.

EMS memory is not as fast as conventional memory. Thus, the use of EMS memory can affect the speed of your program. This difference in speed is especially evident in Lotus 1-2-3, Release 3, and Symphony, Releases 2 and 2.2, where your programs will execute approximately 35% slower if you are using expanded memory.

Lotus claims that you can utilize up to four megabytes of expanded memory. However, this statement is misleading. For each "segment" ("packet") of four active cells that are stored in expanded memory, Lotus allocates about 16 bytes of conventional memory (i.e., ***about*** 4 bytes of conventional memory are consumed for each EMS cell). This conventional memory is used for control purposes and establishes a link to the expanded memory location where each cell's information is located. The amount of unused conventional memory, therefore, will place limitations on the number of cells of data that your Lotus program can support (no matter how much memory each cell takes up in expanded memory).

To summarize: EMS memory can help you handle massive amounts of data, but EMS memory is not nearly as fast as standard memory and it will slow down your spreadsheet and programs. ***Therefore, unless you run out of memory, I recommend disabling expanded memory.*** Your spreadsheets will run about 25% to 35% faster for nearly all operations. You can disable your expanded memory by removing the command lines in the CONFIG.SYS file that specify the EMS device driver (i.e., the ".SYS" file) and then rebooting your computer. Thus, a good strategy to ensure maximum performance is to create your system using conventional memory only. Then monitor memory usage with the /*W*orksheet *S*ettings (Lotus 1-2-3) or {SERVICES} *S*ettings command and determine whether you need expanded memory. (For a more detailed description on how to use these commands see the section entitled "***Configuring Your System For Efficient Memory Usage***," below.)

There is one last EMS quirk that you need to be aware of. If you are using Lotus with a coprocessor accelerator board (such as the Orchid Technologies PCturbo 286e), Lotus spreadsheets WILL NOT be able to access the personal computer's EMS memory in accelerator mode. If you have added one of these boards to an old IBM PC/XT to try to improve its speed, you will need to choose between having access to the personal computer's EMS memory and speed. However, coprocessor boards that have their EMS memory on daughter boards (such as the Intel Inboard 386/PC) will allow Lotus to access EMS memory on the daughter boards.

Extended Memory. Extended memory (also known as "Protected Mode Memory") is memory above 1024 kilobytes (1 megabyte). The 80286 and 80386 microprocessors can access extended memory directly. The 8088 microprocessor cannot.

Only certain programs can utilize extended memory. Currently in the Lotus family of products, only Lotus 1-2-3, Release 3, can utilize extended memory. In fact, Release 3 **requires** extended memory. A stock PC or XT computer without a coprocessor board cannot access extended memory, because they have 8088 microprocessors. It is for this reason that Release 3 cannot be used on a PC or XT computer unless a coprocessor board has been installed.

Extended memory has two very valuable virtues. The first virtue is that extended memory is much faster than expanded memory. This feature makes program execution and spreadsheet recalculation much faster. Furthermore, unlike expanded memory, extended memory does not require an allocation of conventional memory for control purposes. Thus, the maximum cell limits described in the preceding section would be virtually eliminated with extended memory when new releases of Lotus utilizing extended memory are made available.

Local Area Network. For most database management applications described in this book, a single personal computer is sufficient. A single personal computer can handle most LIM system requirements for a single laboratory and can be used to issue written reports to other laboratories, managers, etc.

If you need to share your data with other terminals, you will need a Local Area Network (LAN). A LAN is an group of stations (computers) interconnected by communications channels that include special plug-in cards, wiring, etc.

There is a wide diversity of LANs available. Implementation and maintenance of a LAN can be quite intricate and is dependent on the type of LAN being used. For this reason, LAN specifics will be considered outside the scope of this book and will be discussed only as an interface to the database being managed by the Lotus LIM system.

Tape Backup. LIM system databases containing information on thousands of samples represent a considerable time investment. Consider the value of these records and how long it would take to recreate them. It could take months re-keying in 30 megabytes of records and cost thousands of dollars in clerical time, ***only if*** the information were available in the first place. You must protect your investment. The best protection is to use a tape backup system. The price of tape backups has plummeted (<$500), making them very worthwhile investments.

Once installed, most tape backup systems provide an automatic backup feature. With this feature, the software automatically makes backups of specified data files during quiet hours. For example, you can set the system to backup automatically during lunch time or at midnight, provided the computer is free.

Tape backups are available from many vendors. The system I use is very reliable and fast. The system is from inmac® Corporation (2465 Augustine Drive, Santa Clara, CA 95054-9977).

CONFIGURING YOUR SYSTEM FOR EFFICIENT MEMORY USAGE

Both @BASE and your Lotus spreadsheets are stored in your computer's conventional memory. The @BASE data files are stored on disk. When working with large data files, storing them on disk rather than in memory allows you to store many more data records.

However, @BASE and the @BASE Option Pac add-in application programs occupy substantial amounts of conventional memory, limiting the size of the spreadsheet you can create before you obtain a "Memory Full" message. This size limitation is especially evident in Lotus 1-2-3, Release 2.2. Therefore, it is important for you to determine the amount of available memory after all of your application programs have been added in, verify that enough conventional memory remains for your spreadsheet, and modify your system accordingly.

To determine the amount of remaining memory in Lotus 1-2-3, issue the /*W*orksheet *S*tatus command. In Symphony, the equivalent command is {SERVICES} *S*ettings. Lotus will respond by providing you with a display of the current status of the system. Under "Conventional memory," the first number indicates how much memory is currently available. The second number indicates the total available. When you attach @BASE and the @BASE Option Pac, the total available decreases. In many cases, this decrease results in an unacceptably low amount of remaining memory. For this reason, you must configure your system to make more conventional memory available. The following is a list of options that you can use to retrieve memory.

- The DOC, FORM, and COMM environments of Symphony are not normally needed and can be unloaded with a program called "EXTRA K". This program is available from Lerman Associates (12 Endmoor Road, Westford, MA, 01886, 508-692-7600) or from the author using the order form at the end of this book. Removing these three environments with EXTRA K will provide up to 3,000 more cells for spreadsheet use in systems without expanded memory and up to 21,000 more cells in systems with expanded memory.
- The Undo feature of Lotus 1-2-3, Release 2.2, requires a substantial amount of memory (about 163,400 bytes). In fact, this amount of memory is so large that you will get a "Memory full" message immediately after @BASE and @BASE Option Pac are loaded. It is therefore important to disable the Undo feature with Release 2.2. To do so, issue the /*W*orksheet *G*lobal *D*efault *O*ther *U*ndo *D*isable *U*pdate *Q*uit command. By doing so, you will free about 6,000 cells worth of memory in systems with only conventional memory and 42,000 cells in systems

with expanded memory. Upon returning to the spreadsheet, you will note that the "Memory full" message will still be lit. This message indicates that all of the memory available has not been released back to the pool. To release the memory, exit and re-enter Lotus 1-2-3.

- If you have Lotus 1-2-3, start the program with "123" instead of "LOTUS". If you have Symphony, start the program with "SYMPHONY" instead of "ACCESS". By doing so, you will conserve about 6,500 bytes of conventional memory.
- Remove any memory-resident (Terminate-Stay Resident, TSR) programs not in use. Programs such as SmartNotes and Sidekick are loaded into memory before Lotus is loaded, leaving less room for spreadsheets. Unloading these programs before entering Lotus will make more memory available.
- Eliminate any add-in application programs that are not in use, especially the DOS.APP application program for Symphony. (DOS.APP is a particularly memory-intensive program.)
- Eliminate unnecessary data in spreadsheets. If possible, delete rows and columns. Just erasing cells only removes contents, but does not release memory set aside to access the cell.
- Do not format cells that you do not plan to use. Formatted cells are considered to be in use even if they do not contain data and Lotus assigns memory to the cells. To remove formatting from cells, issue the /***R***ange ***F***ormat ***R***eset command. You will recover about 16 bytes of memory for each cell you reset.
- Delete any windows not in use. Removing unneeded windows in Symphony will make more conventional memory available.
- Convert formulae to their values or use macro programs to place values in cells. Formulae in cells take up more memory than labels and numbers. Formulae also slow the recalculation speed of your spreadsheet. If you have used formulae to generate values that do not change as you use the spreadsheet, convert them to their values by issuing the /***R***ange ***V***alues command.
- Use default settings whenever possible.
- Look for cells that contain "null" labels--label prefixes without any text or label prefixes followed by blank spaces. These cells consume the same amount of memory as any other label. Erase or delete these cells and you will recover about 6 bytes of memory for each cell removed.
- Blank cells, formatted cells, and cells containing null labels are very hard to find. Unless you know where to look, it will take a lot of time finding these cells. For this reason, you should avoid formatting empty cells and making null label entries.
- After you have implemented any of the above memory saving suggestions, save the file to disk and retrieve the file to recover the memory. As Lotus saves the spreadsheet, it frees up the memory that had been previously allocated to those entries. When the spreadsheet is reloaded, that memory will be available for new entries.

If, after all of the above suggestions have been implemented, your application is still low on memory, you should enable your EMS memory. If you are using Lotus 1-2-3 with EMS memory, I would recommend investigating a product called "BEYOND 640," available from Intex Solutions, Inc. (161 Highland Ave, Needham, MA 02194, (617) 449-6222.) BEYOND 640 circumvents the conventional memory allocation problem (noted in the section on EMS memory above) that imposes a limit on the amount of EMS memory available. BEYOND 640 circumvents the problem by automatically moving the ENTIRE spreadsheet (including cell address records) to EMS memory when necessary. Thus, BEYOND 640 lets your spreadsheets use the entire four megabytes of EMS memory. Perhaps more importantly, BEYOND 640 prevents EMS memory from being used until a threshold of conventional memory remains. For example, BEYOND 640 can be set to begin using EMS memory when there are 16 kilobytes of conventional memory remaining. When the amount of free conventional memory goes below this threshold, BEYOND 640 starts to store cell address records in expanded memory. This threshold strategy automatically delays using the slower EMS memory as long as possible.

A SIMPLE EXPLANATION OF LOTUS, MACRO COMMANDS, AND @FUNCTIONS

Lotus 1-2-3 and Symphony

Lotus 1-2-3 and Symphony both have over 600 commands. This magnitude tends to overwhelm and discourage new users. What new users fail to understand is that they need to know only a handful of commands to get started.

Consider all of the programming commands available as if they were tools in a toolbox. To perform a certain task, look up the name of the tool and use it. The more commands that you are aware of, the more tools you have at your disposal. ***Half the battle of programming is knowing what is possible.*** To get a grasp of the available Lotus tools, skim the manuals accompanying the Lotus software, browse through Appendices A through C at the end of this book, or look over some of the books listed as references at the end of this chapter.

A number of the most powerful commands are not documented in the Lotus manuals and many of the reference books listed. This book uses many of these undocumented features of the Lotus programming language. It also describes many ways to use the commands in programs...in both conventional and unconventional manners.

Macro Commands and @Functions

A set of instructions in a format and language that Lotus can understand is called a ***Lotus macro program***. The language that is used is called the ***Lotus Command Language*** and the instructions are called ***macro commands***.

There are six broad categories of Lotus macro commands. They are

- *System commands*, which allow you to control the screen display and computer's speaker.

- ***Interaction commands***, which allow you to create interactive macros that pause for the user to enter data from the keyboard.
- ***Program flow commands***, which let you include branching and looping in your program.
- ***Cell commands***, which transfer data between specified cells and/or change the values of cells.
- ***File commands***, which work with data in DOS files.
- ***Menu commands***, which allow you to design and manage menus.

Special-purpose Lotus formulae are called ***@functions***. Lotus contains numerous @functions. Each @function can extend your calculating power beyond simple arithmetic and text handling operations. There are several broad categories of Lotus @functions: mathematical, text, logical, financial, statistical, database, and date/time.

@BASE and the @BASE Option Pac also supply sets of @functions, called ***@DB*** and ***@NDX @functions***. The @BASE @functions establish an active bi-directional link between disk stored database files and worksheets. The @NDX @functions communicate with index file data to establish efficient connections between the spreadsheet and database. The @NDX @functions are primarily used to locate record numbers. Once record numbers are located, @DB @functions are used to analyze, perform calculations, and update databases from within the spreadsheet while the database files remain on disk. This book makes extensive use of @BASE @functions to open database files, perform calculations on records, locate specific records, modify fields, close database files, etc.

In general, macro programs use both macro commands and @functions to carry out the kinds of processes that your program implements to perform database chores. Your goal is to plan and write efficient macro programs. Efficient programs

- get specific work accomplished in a specific order.
- specify the exact functions that need to be carried out and when.
- manipulate data in the cells of your spreadsheet.
- manipulate data in fields of your databases.

Planning ahead is perhaps the most important concept of efficient programming. Knowing what you need to do, what you can do, and when you need to do it is over 90% of the program development game.

YOUR KEYS TO SUCCESS

If you work through this entire book, you should be well equipped to create your own database management applications. The real secret to becoming an expert, however, is to actually work through the examples contained within this book. You will be able to get a much better feel for the structure of a program if you type the examples into a spreadsheet. Hands-on experience is the best way for you to learn. More importantly, if you have experimental data that can be maneuvered using a method being described in a chapter, you should use the data to learn more about the technique.

After you have a general idea of what each example template or Lotus program can do, you should work with any features that you don't fully understand. Do so by making variations to see the effects of your changes. Place a copy of the formulae that are within Lotus program commands into cells of the spreadsheet and see how they update as the program proceeds. Frequently, after you try using a specific Lotus command or formula and see it in action, the explanation in the book will become clearer and you will learn something important about the database management concept.

One thing that you will discover when reading through this book is that there are often many solutions to a single problem. There often isn't a single, best way to achieve a programming goal with software. I will usually present only one or two ways to address each programming challenge. However, with a programming language as versatile as this one, there are usually ***many*** ways to implement almost any task. Some ways are better than others, depending on the circumstances and the application. Sometimes you will be surprised to find that what you thought was a very inefficient method, is actually the most efficient. Do not be afraid to investigate several alternative programming schemes to accomplish a particular task and then choose the best.

DISCLAIMER

Lotus and Personics will give you no guarantees and neither can I. However, I can assure you that I have made a strong effort to ensure that all of the programs that I present to you in this book work. I have tried all of them in both Lotus 1-2-3 and Symphony on an IBM® XT™, AT®, and Model 80. However, the examples are just examples. Although all of the examples can be used directly in your laboratory, they are meant to teach you certain specific concepts about how to create database management tools and are not necessarily the most efficient way to program for your particular situation.

SUMMARY OF WHAT HAS BEEN ACCOMPLISHED

This chapter provided basic definitions for terms that will be used throughout the remainder of this book. For the purposes of this book:

- A ***field*** is an item of information.
- A ***record*** is a collection of related information (fields).
- A ***database*** is a collection of related records.
- An ***index*** is a file that contains information about the location of records in a database file.
- An ***extended relational database*** is a group of two or more related files that are joined and work together so that information can be shared among the files.
- A ***Laboratory Information Management System (LIMS)*** is a database specifically tailored to an individual laboratory's needs.

This book will teach you how to create a system that will provide the successful storage and retrieval of informational and experimental data. To ensure that you have the necessary software and equipment, the chapter provided a catalog of items that you will need before you begin to create your database system.

WHAT'S NEXT?

The next chapter will provide you with a description of the example database system that will be used throughout the remainder of the book. More importantly, the chapter will explain how to design your database LIM system to fit your needs.

Beginner Users' References

ADAMIS, EDDIE, *Command Performance: Lotus 1-2-3.* Bellevue, WA.: Microsoft Press, Div. of Microsoft Corp., 1986.

MEZEI, LOUIS, *Laboratory Lotus®: A Complete Guide To Instrument Interfacing.* Englewood Cliffs, NJ.: Prentice Hall, Inc., 1989.

MEZEI, LOUIS, *Prentice Hall Laboratory Lotus® Series™: Practical Guide to Statistics and Curve Fitting for Scientists and Engineers.* Englewood Cliffs, NJ.: Prentice Hall, Inc., 1989.

MILLER, STEVEN, *Lotus Magazine: The Good Ideas Book.* Reading, MA.: Addison-Wesley Publishing Co., Inc., 1988.

The staff of Lotus Books, *The Lotus Guide To Learning Symphony Macros.* Reading, MA.: Addison-Wesley Publishing Co., Inc., 1986.

The staff of Lotus Books, *The Lotus Guide To Learning Symphony Command Language.* Reading, MA.: Addison-Wesley Publishing Co., Inc., 1985.

STARK, ROBIN, *Encyclopedia of Lotus 1-2-3.* Blue Ridge Summit, PA.: Tab Books, Inc., 1987.

Advanced Users' References

FENN, DARIEN, *Symphony Macros And The Command Language.* Indianapolis, IN.: Que Corporation, 1987.

ZUCKERMAN, MARCIA, DYRUD, ANNE, and POSNER, JOHN, *The Lotus Guide to 1-2-3 Advanced Macro Commands.* Reading, MA.: Addison-Wesley Publishing Co., Inc., 1986.

CAMPBELL, MARY, *1-2-3 Power User's Guide.* Berkeley, CA.: Osborne McGraw-Hill, Inc., 1988.

Database and LIMS References

ANDERSEN, DICK, and WEIL, BILL, *1-2-3 Database Techniques.* Carmel, IN.: Que Corporation, 1989.

MCDOWALL, R.D., *Laboratory Information Management Systems: Concepts, Integration, and Implementation.* Wilmslow, United Kingdom: Sigma Press, 1987.

SIMPSON, ALAN, and MOSICH, DONNA, *Focus on Symphony Databases.* Berkeley, CA.: Sybex, Inc., 1986.

2

System Design

The first, and most important, step in creating your information management system is to determine what information you want to keep track of. In a very real sense, the material provided in this chapter is the most important in the entire book. Once you have defined the information you want your database system to contain, building an information management system is trivial. Modifying an existing database to accept more fields is also trivial. But adding information to an existing database containing 10,000 or more records is another matter. Unavailability of historic data to update a field of information from the date that the system was first implemented is the worst scenario. Even with the best scenario, typing thousands of datum into an existing database to update the database to include a new information field is very tedious to say the least ... and the chances of introducing a whole new round of mistakes and software bugs make modification even more unpleasant.

So, whether you are creating a database to simply archive data from a single laboratory or assay or are creating a fully functional Laboratory Information Management system, take the time to plan. Take it from me, the time you take now will be very well spent and will save time later. This chapter will give you some guidelines that have proven effective for the systems I have created.

DESIGN COMMITTEES

Get as many people involved in the design process as you can. Form a committee of *everyone* that will use the system or receive reports generated by the system. When forming the committee, do not forget that both management and technician-level personnel also need to be involved. Management will benefit greatly from reports on the number of tests, cost per test, backlog, average turnaround time, departmental charges, etc., and their input is crucial. Technician-level personnel will help ensure ease of use for users and minimal training for new hires.

Meet at least weekly for a couple of weeks. Keep the meetings short. Optimally, one-hour meetings are best after the preliminary meeting. Issue minutes. Once everyone has given their input, prepare a document containing the specifications for the system. Circulate the document for comment. Hold meetings to review the comments. Issue final specifications and begin to create the system only after consensus has been reached.

Once you have created the system, consider the system open to review and modification for the first few weeks. The objective of this review stage is to catch problems as early as possible and fix them *before* your database becomes too unwieldy or large to allow easy update. Programming errors will inevitably surface and need to be fixed. Missed database fields will also become apparent. Therefore, you need to hold weekly design review meetings to allow input. You also need to review the changes made so that everyone understands the ramifications. These design review meetings often reveal problems and conflicts that can arise from changes that benefit one department while impairing another.

Everyone involved needs to continue with their manual record-keeping during this period. Written records may seem redundant, but the backup will provide a high degree of safety in case a major problem occurs that destroys data in the database. Written records also provide a convenient way to check system accuracy. Side-by-side comparisons are the best way to check the database. Each individual attending the design review meeting should compare their manual data with reports issued to them by the system. The design review meeting is an ideal forum to expose interdepartmental data integrity problems that are often difficult to uncover independently.

FIRST LEVEL LISTS

A good place to start designing your system is to think about what you are trying to achieve by implementing the system. Begin by reviewing and broadly defining your goals and objectives for putting in a LIM system and what you want to get out of it. Write the goals and objectives down on a white board or flip chart. Then start to determine how you plan to use the information and what kinds of information will be needed to meet your objectives. The kinds of information will specify the data you will need to keep. The data specifications will define the fields you will need for the database.

The following are some of the first level lists you will need:

- A list of the advantages or benefits of a LIM system to your organization.
- A list of the written reports that you want generated.
- A list of the reports you want displayed.
- A list of the ways that you want to search (query) the database.
- A list of the graphics you want generated.
- A list of the types of assays you will be performing.

Benefits Summary

Increased Productivity
Improved Data Integrity
Easy Data Manipulation
Increased Management Awareness
Minimized Paperwork
Easy Trend Analysis
Rapid Reporting
Low-Cost, Sample Archiving And Retrieval
Improved Communications Between Lab, Management,
and Other Departments

Figure 2-1

As you can tell, the lists you generate bracket the operation of the LIM system. That is, you are deciding on what you expect to achieve by the database and what you expect to get out of the database.

The next subsections in this chapter describe the first level lists in more detail and show you how to think about each list so that you can use the lists to identify database fields needed to support the items on the lists. Thereafter, sections describe how to define the field names, field types (string, numeric, date, computed, etc.), number of decimals (for numeric fields), and field widths. The final sections illustrate how to use first level lists to create specifications for searching and joining databases.

Advantages and Benefits of a LIM System

The "Advantages and Benefits Summary" is actually an overview of what you are trying to achieve. The intent for creating this summary is to provide a springboard to defining what kinds of reports you will need to achieve your goals. Once the reports are defined, they will, in turn, lead to defining some of the fields required. Figure 2-1 shows some benefits that should be applicable to most laboratories. The following points are some examples of how to translate your Benefits Summary into reports:

- With "Increased Productivity," you should think about the kinds of reports your LIM system can provide to achieve your goal. For example, you may want to issue work lists to individuals specifying tests to be completed, priority lists, instrumentation usage tallies, etc. Instrument utilization can help identify whether new equipment is needed and whether the equipment should be purchased or leased.
- "Data Integrity" should start you thinking about

- whether you want to create databases containing setup information for instruments. (For example, the string commands necessary to configure instruments through their RS-232 ports can be stored in database files and retrieved when needed.)
- how you want the flow of information to take place between instrument, spreadsheet, database, and final report.
- how to test user input for correctness, evaluate results from controls to ensure that they are within specified limits, etc.

- Your LIM system can also provide reports to "Increase Management Awareness". These reports can contain information on number of samples run within a period of time, on-time performance, trends, cost per test, etc. The reports can be used to justify more instrumentation and/or people, create departmental budgets, etc. Lists can also be generated to expose departments that order a high volume of tests, so justification can be evaluated.
- Under "Easy Trend Analysis," reports and graphics may be specified to study the day-to-day results of controls (e.g., "Quality Control Charting"). Trend Analysis may also contain data for determination of the stability of a material.
- The "Low-Cost, Sample Archiving and Retrieval" benefit is included to help you remember to begin thinking about how you want to search for data. Pinpointing the correct data requires fields that can be used for the search. You should think about the search fields that you will need. The sample list should also prompt you to think about what kinds of reports you want to put out on each sample and the information that needs to be included on each report. For example, do you want a simple report to contain just a group of samples by barcode numbers and results, or do you want individual, formal Certificates of Analysis on each sample? Do you want all samples from a single department to appear together on a single report? Do you want a list of the status of samples, the samples' storage locations, etc.? Do you want automatic billing?

Once completed and defined, your "Advantages and Benefits" list will lead to the next and most important list ... the "Reports" list.

Reports to be Generated

Before creating a database to achieve the goals defined in the previous section, you must decide on how you plan to use the data. The best way to do this planning is to start with the output. The output may be a hard-copy report that you want to print for distribution or just some selected information that you want to view on the display. When creating the "Reports To Be Generated" list, think about what information you want to look up and how you will look it up. Once you know what information you plan to get out of a database, you will know what information to put into it. One aid is to actually sketch out the reports you plan to issue. From the sketches, you will begin to identify some of the database fields containing the information that will be needed for the reports and fields

Reports to be Generated

Sample Receipt
Sample Location
Status Reports by Sample
Status Reports by Test
Overdue Samples
Worklists for Laboratory Staff
Test Results by Sample
Results of All Samples in Assay Batch
Certificates of Analysis
Backlog
Reagent Utilization
Instrument Utilization
Trend Analysis by Test
Quality Control Results by Test
Results by Worklist
Statistics
Approval and Release
Invoice Report by Sample
Customer Billing
Cost Per Test
End of Year Summary

Figure 2-2

required for searches to pinpoint the data. For example, suppose you wanted a report containing how many tests were completed by a laboratory over a period of time. Minimally, you would need fields for specifying date run and sample identification. However, you may want to break the report down into categories such as type of assay, the date a sample was received, the date the report was issued, etc. You may even want to issue barcode numbers of samples not yet processed. Another example would be a report on trend analysis. For trend analysis you may need fields for date run, assay type, sample identification, results, technician, etc. Some of the fields would be used to locate information (for a certain test type and within a specified time period, for example) and other fields would be used for data to be presented (e.g., results, number of tests performed by technician, etc.)

By designing the reports you want and the criteria needed to locate the information required to generate the reports, you will

- be more efficient at not missing fields of information.

Graphs to be Generated

Stability of Material
Standard Curves
Quality Control Charts
Tests Requested by Department
Tests Requested by Product Group
Total Tests Requested by Month
Cost Per Test by Month
Backlog Trends
Status of Samples by Project Code
Productivity by Technician

Figure 2-3

- reveal fields that need to occur in files that you join together when creating a relational database.
- reveal fields that can be used to retrieve groups of data that need to be accessed during a database search (e.g., data from all samples run in a single assay batch).

Figure 2-2 shows some typical reports. As you might expect, the number of reports that can be generated can be very large. If hard copy reports are different than displayed reports, you should create separate lists. As you create the lists, go back and forth between the "Reports" list and "Benefits" list created in the previous section to make sure you have not missed anything.

Graphics to be Generated

Not surprisingly, "Graphics" lists are similar to "Reports" lists. After all, graphics are essentially reports. Figure 2-3 illustrates some possible graphs that can be generated. As in the case of "Reports," you should begin thinking about the fields that would be necessary to search for the information you want from the database. You should also begin thinking about how you want to format the data being placed into a spreadsheet being used by Lotus to generate the graphics.

Types of Assays to be Performed

At first glance, the list of "Types of Assays To Be Performed" appears to have limited utility. However, by defining full and valid acronym names for all of the analyses that you plan to perform, you can provide a number of features that make your LIM more

Types of Assays

Tumor Necrosis Factor (TNF)
Macrophage-Colony Stimulating Factor (M-CSF)
Platelet-Derived Growth Factor A (PDGF-A)
Platelet-Derived Growth Factor B (PDGF-B)
Interferon (IFN)
Interleukin-1α (IL-1A)
Interleukin-1β (IL-1B)
Interleukin-2 (IL-2)
Deoxyribonucleic Acid (DNA)
Type β Platelet-Derived Growth Factor Receptor (PDGF-R)

Figure 2-4

reliable. For example, you can use the full and acronym names to design and create

- user input validation systems.
- menus with choices corresponding to each test to be performed.
- macros to retrieve appropriate spreadsheets for different test formats.
- a variety of spreadsheet data acquistion and analysis macro programs that execute depending on assay type.
- database lists of instrument configuration settings for each assay.
- databases of assay protocols.
- cost per test files for billing.
- systems to ensure successful searches.

This list is based on the fact that your LIM system needs to be designed to accommodate several users. You therefore need a reliable way to ensure that everyone using the system specifies the correct assay names during input and searches. For example, to various users, the acronym for Platelet-Derived Growth Factor A may be "P-DGFA", "PDGF-A", P-DGF-A", or "PDGFA". If a database contained four (or more) different entries for the same test type, it would be difficult (if not impossible) to design a reliable system to search for data pertaining to the assay type. However, if you prepare a list of valid test types and acronyms, you can create menu systems that execute macro programs that, in turn, use only a limited number of specifically defined names for the tests. These macro programs would then input reliable test names in the database and later search the database for the names. This system ensures reliable searches without the user having to guess the name of the test in the field.

Search Criteria

Sample Description by Barcode Number
Sample Results by Barcode Number
Sample Results by Barcode Number and Assay Type
Sample Results by Assay Type, Date, and Batch
Batch Results by Assay Type, Date, and Batch
Samples by Status and Request Date
Samples by Type of Request and Status
Reagents by Lot Number
Standards by Lot Number
Controls by Lot Number
Components by Lot Number

Figure 2-5

Search Criteria

In a fully implemented LIM system, there may be thousands of records. You generally only want to work with a small subset of these records. To choose only the records you are interested in on a given occasion, you can specify criteria conditions for a file. ***Criteria*** are a means of selecting specified portions of your database for @BASE processing. Records are selected by telling @BASE to include only those records that meet the specified conditions (criteria). Records that do not meet the criteria are temporarily ignored.

Criteria specifications include two pieces of information: the conditions that need to be met and the field to be evaluated. Thus, you need to include fields in your databases and index files that will allow you to find the information you are looking for. The "Search Criteria" list will help you.

For example, suppose you were looking for changes in a sample over time to determine the precision of an assay, spot trends, evaluate the stability of samples, etc. To obtain the information you need, you will need to have sample identification, assay type, and date fields that can be searched and compared to appropriate criteria. That is, when you create your search criteria list you should decide on why you want to keep track of the data and what you plan to evaluate. This decision will give you specifications for how your database needs to be searched. Search specifications will lead to the fields you need in each record. The other fields in the records are the data you need to retrieve.

Perhaps more importantly, the Search Criteria list will lead to the fields you need in index files. If you recall from Chapter 1, index files substantially increase the speed at which data can be found in databases. By building index files with search criteria fields

Reagents Database (REAGENTS.DBF)

Identification Fields

```
Record Number (from @BASE)                          Recno:N:8:0
Lot number of entire set                            LOT_RGNT:C:20
```

Information Fields

```
Reagent name                                        RGNT_NAME:C:20
Reagent description                                 RGNT_DESC:C:25
Date of preparation                                 R_PREP_DT:D:9
Shelf life                                          R_SHELF_LF:N:8:0
Expiration date                                     R_EXP_DT:E:8
Expiration flag                                     R_EXP_FLG:E:7
Lot number of component #1                          LOT_R1:C:20
Lot number of component #2                          LOT_R2:C:20
Lot number of component #3                          LOT_R3:C:20
Lot number of component #4                          LOT_R4:C:20
Technician that prepared reagents                   RGNT_TECH:C:5
```

Figure 2-6

in them, you can achieve much faster searches and save time.

Figure 2-5 shows a typical Search Criteria list. In reality, searches are most often performed on multiple criteria working together to pinpoint appropriate records. Your "Search Criteria List" will help you identify those fields that are not necessarily displayed in reports, yet important in the identification of data and need to be included in databases being searched.

FIELD SPECIFICATIONS

For each field in a new database, the following specifications are needed:

- The name of the field.
- The field type (character, numeric, logical, or date).
- The field length (number of characters allowed).
- The number of decimal places (if field is numeric).

Figures 2-6 through 2-9 show some typical specifications and illustrate a set of example fields that would be required to describe a portion of a relational database for a LIM system that keeps track of information for a biotechnology laboratory. For the moment, a focus on the details of data distribution between databases will be set aside and focus will instead concentrate on field definitions only.

Standards Database (STANDARD.DBF)

Identification Fields

Record number (from @BASE)	Recno:N:8:0
Lot number of entire std. set	LOT_STD:C:20

Information Fields

Name of standard set	STD_NAME:C:20
Standard set description	STD_DESC:C:25
Date of preparation	S_PREP_DT:D:9
Shelf life	S_SHELF_LF:N:8:0
Expiration date	S_EXP_DT:E:8
Expiration flag	S_EXP_FLG:E:7
Lot number of standard #0 (Blank)	LOT_S0:C:20
Concentration of standard #0	CONC_S0:N:10:3
Lot number of standard #1	LOT_S1:C:20
Concentration of standard #1	CONC_S1:N:10:3
Lot number of standard #2	LOT_S2:C:20
Concentration of standard #2	CONC_S2:N:10:3
Lot number of standard #3	LOT_S3:C:20
Concentration of standard #3	CONC_S3:N:10:3
Lot number of standard #4	LOT_S4:C:20
Concentration of standard #4	CONC_S4:N:10:3
Lot number of standard #5	LOT_S5:C:20
Concentration of standard #5	CONC_S5:N:10:3
Lot number of standard #6	LOT_S6:C:20
Concentration of standard #6	CONC_S6:N:10:3
Lot number of standard #7	LOT_S7:C:20
Concentration of standard #7	CONC_S7:N:10:3
Technician that prepared standards	STD_TECH:C:5

Figure 2-7

Field Names

Each field name acronym can be up to 10 characters in length. However, you should try to keep field names that will be used in indexes and joined fields as short as possible.

Try to use names that are as descriptive of the data being held in the fields as possible. Legal characters for field names are: letters of the alphabet, integers 0 through 9, and the underscore character ("_"). The first character of a field name cannot be an integer and field names cannot contain spaces. You do not have to use upper-case letters, @BASE will convert to upper-case automatically. Therefore, you may type either assay_type or ASSAY_TYPE.

The following rules apply to field names:

- Field names within a file ***must be unique***.

Assay Batch Database (BATCH.DBF)

Identification Fields

Record number (from @BASE)	Recno:N:8:0
Assay type	ASSAY_TYPE:C:10
Date assay performed	RUN_DT:N:5:0
Batch number	BATCH:N:3:0
Lot number of standards used	LOT_STD:C:20
Lot number of reagents used	LOT_RGNT:C:20

Information Fields

Technician performing assay	BCH_TECH:C:5
Average response for standard #0	AVG_S0:N:8:4
Average response for standard #1	AVG_S1:N:8:4
Average response for standard #2	AVG_S2:N:8:4
Average response for standard #3	AVG_S3:N:8:4
Average response for standard #4	AVG_S4:N:8:4
Average response for standard #5	AVG_S5:N:8:4
Average response for standard #6	AVG_S6:N:8:4
Average response for standard #7	AVG_S7:N:8:4
Curve fit-A Parameter/Slope	CURV_A:N:8:4
Curve fit-B Parameter/Intercept	CURV_B:N:8:4
Curve fit-C Parameter	CURV_C:N:8:4
Curve fit-D Parameter	CURV_D:N:8:4
Standard Error of Y Estimate	STD_Y:N:8:4
Correlation Coefficient Squared	R_SQR:N:6:4
Number of Observations	N_OBS:N:3:0
Degrees of Freedom	DF:N:3:0
Standard Error of Coefficient	STD_C:N:7:4
Average response for control #1	AVG_C1:N:8:4
Average response for control #2	AVG_C2:N:8:4
Date of final review by supervisor	FREV_DTE:D:9
Technician comments	TECH_COMS:C:21

Figure 2-8

- Field names used to create links between two files when they are joined ***must be identical*** in both files.
- Names of lookup fields selected for viewing as a result of a join ***must be different*** than all of the field names existing in the primary file obtaining the information and any lookup fields from any other file joined to the primary file.
- Field names in macro programs, database files, and index files that function as specifications for data to be transferred between databases and spreadsheet templates or other databases, ***must be identical***. For example, ASSAYTYPE and ASSAY_TYPE would be treated as two completely different entities by both Lotus and @BASE. If a macro program tried to access ASSAY_TYPE in a database file by calling it ASSAYTYPE, an error message would be issued.

To ease compliance with the above rules, you should organize your fields into two

Samples Database (SAMPLES.DBF)

Identification Fields

Field	Definition
Record number (from @BASE)	Recno:N:8:0
Sample barcode number	BARCODE:N:8:0

Information Fields

Field	Definition
Sample name/description	SAMP_NAME:C:20
Type of assay(s) requested	TYPE_REQ:C:32
Requester's Name	REQ_NAME:C:10
Date of request	REQ_DT:D:9
Requester's notes and comments	RQS_NTS:C:30
Department number	DEPT_NUM:N:3:0
Project code	PROJ_CDE:C:6
Date sample will be delivered	SMP_DEL_DT:D:9
Sample rack barcode number	RACK_BCD:N:8:0
Storage location	STOR_LOC:C:10
Sample Status	S_STATUS:C:4

Assay Results Database (RESULTS.DBF)

Identification Fields

Field	Definition
Record number (from @BASE)	Recno:N:8:0
Sample barcode number	BARCODE:N:8:0
Type of assay performed	ASSAY_TYPE:C:10
Date assay performed	RUN_DT:N:5:0
Batch number	BATCH:N:3:0

Information Fields

Field	Definition
Plate number	PLATE_NUM:N:8:0
Dilution Factor	DIL_FCTR:N:10:0
Raw data point #1	RAW_DATA1:N:8:4
Raw data point #2	RAW_DATA2:N:8:4
Results to be reported	RESULTS:N:10:3
Results Status	R_STATUS:C:4

Figure 2-9

categories and assign names according to category:

- ***Identification Fields:*** Fields that will be used as a basis for searching, sorting, and joining databases. The Identification Fields section can be thought of as the bookkeeping portion of a database file.
- ***Information Fields:*** Fields that will be used for archiving data for later retrieval. The Information Fields section can be thought of as the record-keeping or data archiving portion of a database file.

Note that Figures 2-6 through 2-9 are organized according to these two categories. The following are some examples in the figures that illustrate how field organization affects the names applied to fields:

- ***Field names used to create links between two files when they are joined must be identical in both files:*** Illustrations of this concept are exhibited by the LOT_RGNT and LOT_STD field names in the Identification Fields sections of the REAGENTS and STANDARD database files of Figures 2-6 and 2-7, respectively. Correspondingly identical names of these two fields appear in the Identification Fields section of the BATCH.DBF file of Figure 2-8. These common field names provide the links between BATCH and the other two files when the files are joined to create an extended relational database.

  Another good example of this rule is shown in the field names of Figures 2-8 and 2-9. When a large number of samples are to be performed during a day, the samples are often divided into smaller groups and the assay is performed on each group. This technique is called ***batching***. If the assay is performed in microtiter plates, 96 individual assays can be performed in each plate and a batch can consist of several plates. Several batches may be run on a single date. One way to fully characterize a batch is to create fields called ASSAY_TYPE, RUN_DT, and BATCH. Together, this trio of fields eliminates the need to keep a continually updated list of all batches run from the inception of the system just to assign a number to a batch. That is, the technician needs only to enter the type of assay, current date, and batch number on that date to achieve uniqueness for a batch. This trio can then be used to efficiently store and retrieve information for each of the many samples in a batch.

  For each sample, a record is entered into a database file (e.g., RESULTS.DBF of Figure 2-9) that contains the trio of fields and any other pertinent information. Next, a record is entered into another file (e.g., BATCH.DBF of Figure 2-8) that also contains this trio of fields, along with fields that describe details that are common to ***all*** of the samples in the batch. When the two database files are joined, the common information in the lookup file (BATCH.DBF) is made available to each record in the primary file (RESULTS.DBF) that contains matching data in the trio of fields. For example, a single entry can be made in a lookup database file that contains details about the standard curve, technician performing the assay, lot number of standards, lot number of the reagents, etc. Thus, a single entry contains the information for all of the samples in the batch. This system is made possible by designing the trio of field names in the BATCH.DBF and RESULTS.DBF database files to be identical.
- ***Names of lookup fields selected for viewing as a result of a join must be different than all of the field names existing in the primary file obtaining the information:*** The first letter or two of Information Field names in the figures have been coded to the database name. This technique is especially important if several lookup files will be joined to one primary file. Suppose, for example, the REAGENTS and STANDARD databases in Figures 2-6 and 2-7 had identical

field names for date of preparation (e.g., if PREP_DT had been used as a field name in both files). @BASE would not allow PREP_DT to be specified as a virtual field name for any files past the first file joined. However, if PREP_DT had a preceding R_ for the REAGENT database, S_ for the STANDARD, and so on, the rest of the name could be the same and the fields would be available for viewing. This convention also makes it easier to recognize the origin of correponding fields. In Figure 2-7, it would be easy to discern that S_PREP_DT, S_SHELF_LF, S_EXP_DT, and S_EXP_FLG were all fields in the STANDARD database file. This convention makes programming easier and viewing data less confusing.

The LOT_RGNT and LOT_STD field names in the REAGENTS and STANDARD databases illustrate a corollary to the previous paragraphs. Note that each of these field names are unique. From the previous paragraphs, you may have inferred that these names should have been identical. However, the REAGENTS and STANDARD databases will all be joined to the same primary database (BATCH.DBF). BATCH.DBF cannot have redundancy in its field names. Using a field name like "LOT" in all three databases would have created an unacceptable redundancy when an attempt was made to link all three files to BATCH.DBF. Giving the fields unique between-file names created an acceptable environment for linking. Also note that, as explained above, the convention used was designed to make programming easier and viewing data less confusing.

- ***Organize Information Field names into lists:*** The REAGENTS and STANDARD databases illustrate another point about naming Information Fields. Both databases contain lists of fields that hold repetitive data. Each name in the list contains a core acronym, a character relating to the name of the database, and the number of the item in the list. For example, in the STANDARD database, there is a list of lot numbers for each of the sub-components in the set. The field names on the list contain a "LOT_" core acronym, an "S" for standard, and numbers 0 through 7 for each sub-component.

  A similar situation occurs in the BATCH database. This database has three running lists. The first list is for average response of each standard (AVG_S0 through AVG_S7). The second list is for curve fitting parameters (CURV_A through CURV_D). The third list is for average response for each control (AVG_C1 and AVG_C2). By organizing data into lists and naming fields within the lists in an orderly fashion, your databases will be easier to understand and work with.

Field Types

The first specification after the colon in the figures is the field type. Field types can be character, numeric, logical, or date. Following are some specifics:

- ***Character*** fields consist of any alphanumeric characters. If numbers are entered into the field, they are treated as characters and lose their ability to be manipulated mathematically.

- ***Numeric*** fields consist of numbers, decimal places, and +/- signs.
- ***Logical*** fields contain one character, **T** for true or **F** for false.
- ***Date*** fields are displayed in the MM/DD/YY format and are stored in the dBASE III "YYYYMMDD" format. Time is not supported.

Choosing the correct field type is very important. If you change a field type later, the existing data in the field will be converted to the new field type. If the data types are incompatible, the data for that field will be lost. (For conversion tables, see Chapter 4.)

It is also important to have field types consistent between files that are to be joined. In the example figures, all lot number fields are defined as character fields. If two files were joined using lot numbers defined as characters in one database file and numeric in the other database file, the information link would not be effective.

There is a character limit that you cannot exceed when creating index files. (Index files are recommended for searching and sorting, and necessary for joining, databases.) The number of characters in equations for indexing depends on the number of fields being joined, the field types, and the length of the field names. One of the most significant contributors to the consumption of characters is dependent on field type. Following are some details:

- Equations are used by index files to conduct searches or to join files. The equations used and values returned by the equations cannot exceed 100 characters. If you exceed this limit, @BASE will issue an "Index expression too long" message, indicating that the index equation, or resulting string, is too long. This limit appears to be quite large. However, it is not and it is quite possible that you will receive the message, especially if you are using compound key expressions and date fields. If you receive the message, you should consider changing field types. Date fields consume a large number of characters before they can be used in indexing equations. Three functions are required to convert dates to numeric and three functions are required to transform the converted numbers to strings. These six functions add up to a substantial number of characters. However, if a date field is instead defined as numeric (with a field length of five and zero decimal digits) and a Lotus @DATE @function is used to convert the date to a serial number prior to transferring the date to the database, only one function is required. The single function contains substantially fewer characters than the six functions required to convert dates.
- Another reason to keep index equations as small and compact as possible is that an index equation is re-evaluated for each record in the index file. If there are thousands of records in the file, a substantial amount of time may be required. This potential time problem is another reason to avoid using dates for fields to be used in indexes. As explained above, dates require six functions to be executed before they can be converted to usable form. If you build your own user interface, converting ***dates used as keys*** to numbers with the Lotus @DATE

@function prior to storage is highly recommended. If you decide to use the @BASE or dBASE® III user interface, then the choice is a little more complicated. Dates will not be displayed in date format unless the field type is specified to be date. Therefore, user input and review would be impaired because the user would have to work with number equivalents of dates.

- If you find that you must use date fields and/or multiple fields when indexing, keep field names as short as possible. Keeping field names short can sometimes mean the difference between whether the 100-character limit is exceeded or not. For example, suppose that in Figure 2-8 the field names specified in an index file were ASSAY_TYPE, DATE_RUN, and BATCH_NUM and that DATE_RUN was defined as a date field. An "Index expression too long" error message would be obtained during an attempt to create the index file. To preserve the date field type, the names could be changed to ASSAY_TYPE, RUN_DT, and BATCH, respectively. This change would bring the equation to less than 100 characters and would be accepted by @BASE.

To illustrate the use of the above discussion on index design, consider the RUN_DT definition in the BATCH and RESULTS databases. If you note, RUN_DT was defined as numeric in both databases. The RUN_DT fields are used in index files and provide greater efficiency during index searches.

Note in the figures that some of the field types in the figures are listed as "E". The "E" designations mean that the field is not part of the database, but calculated from data in the database using an equation. For example, an expiration date on a reagent (R_EXP_DT) can be obtained by using fields in the REAGENTS database to perform the following calculation:

```
R_PREP_DT + R_SHELF_LF
```

@BASE calls this calculated field a ***computed field***. Computed fields are discussed in more detail in Chapter 5.

Field Lengths

The next specification is the length of the field. Each field can be up to 254 characters in length. With 256 fields per record, each record can have up to 4,000 characters. However, be conservative when defining field lengths. Making very large field lengths for comments is a common temptation. Avoid this temptation. If a comment field length is set to 100 characters, every record created will take up 100 characters worth of disk space. With several thousand records in a database, each with 100-character comment fields, a large amount of disk space can be consumed very quickly.

It is also important to have consistent field lengths between files that are to be joined. In the example figures, the lot number fields in all database files were defined as 20. If two files were joined using a lot number defined as 10 in one database file and 20 in another database file, it would be possible for a problem to occur when labels longer than 10 characters were used to cross-reference information.

When defining lengths of date fields, use a length of at least 9 for spreadsheets. Spreadsheets need a column width of at least 9 spaces to display dates. If your columns are not at least 9 characters wide, asterisks will be displayed instead of dates. @BASE will automatically set date fields to 8, which is appropriate for @BASE displays and database files. You cannot change this specification.

Number of Decimal Digits

The final specification is the number of decimal digits. If you specified numeric as the field type, you can specify up to 15 decimal digits. However, the numeric field length must be at least two digits larger than the number of decimal digits. For example, for 3 decimal digits, the numeric field length must be at least 5.

The STANDARD database in Figure 2-7 illustrates an important point. The standard concentrations have numeric field lengths of 10, and 3 decimal digits. This convention may seem a little odd. However, the database is being set up to allow data for sets of standards for several different types of assays to be archived. Concentrations reported in some assays may be in the 10^6 range, while reports for other assays may be in the 0.001 range. The field length and decimal digit specification cover both ends of the scale and allow for flexibility in the database.

APPORTIONING DATA BETWEEN DATABASES

After all first level lists have been generated, convert the information on the lists into definitions for the fields of information that your databases and spreadsheets need to contain.

Start by developing a preliminary list of database file names. You will need one database file for each group of associated data. For example, if a laboratory offers several different types of biochemical assays for test samples, you would expect to need at least six databases. At least one database would be needed to hold data for each of the following categories:

- Sample registration (background information).
- Sample test results.
- Reagents.
- Controls.
- Standards.
- Batch details.

When deciding on databases, there is a very important DOS limitation that you need to be aware of. The limitation is that although DOS allows up to 255 files to be opened, only 20 files can be accessed by each application. Because @BASE runs within Lotus, the two programs together cannot access more than 20 files, regardless of how many files

are specified in your CONFIG.SYS file. Lotus and @BASE typically have about six open files, leaving about 14 user files remaining. Thus, ***you can have a total of up to about 14 open database and index files before the system begins to issue error messages.*** Because of similarities in controls and standards, a single database is used for both categories of materials and is called "STANDARD.DBF". STANDARD.DBF accommodates both data types and decreases the number of open files by at least two (one database file and one index file). If your system exceeds the 14-file limit, you could decrease the number of files by two more if you merge the REAGENTS and STANDARD databases, because reagents data are similar to standard and control data. However, merging the files will result in slightly more difficult programming.

Next, decide on which data you need to keep track of in each database. As stated in the previous section, there are basically two different types of data fields that databases need to contain. Identification Fields are used as a basis to uniquely identify records in the database and are used for searching, sorting, and joining databases. Information Fields contain data archived for later retrieval. If you recall, Figures 2-6 through 2-9 were divided according to these two definitions.

To begin designing your databases, start by examining your equivalent to each of the first level lists and designating fields that are needed for each database. You will normally need ***one or more*** fields for each of the following categories:

- *Identification Fields*
  - Field(s) that will make each record in a database unique and that can be used for searching the database during queries to unambiguously find information on specified record(s).
  - Field(s) that are used for sorting a database to re-order the information in a specified way.
  - Field(s) that are used for joining databases to achieve extended relational database capabilities.
- *Information Fields*
  - Field(s) for each piece of data to be archived in a record.

- Field(s) that will be used for computing results of other fields and field names for the results.

The first three categories in the above list are fields that are usually included in both database and index files. Although database fields can be used directly for searching and sorting, using index files will increase the speed at which operations will be performed. Index files are required for joining files.

Fields that hold computed results are not material members of a database, but should at least be defined at this time. When designating computed fields, use the letter "E" (for equation) as the data type. (Computed fields will be created in Chapter 5.)

Creating a First Draft

The apportionment of Information Fields between database files is an iterative process. Fields often need to be shuffled from one file to the next before you find the right combination of pigeon holes to achieve optimal system performance and reliability.

A database can have up to 256 fields. The first field specified in Figures 2-6 through 2-9 (***Recno***), is a field for the number of the record in the database and is automatically created by @BASE. So, you should not create the field when you create your database. However, you should include this name so that you remember to include the field in templates that you design.

Thereafter, fields can be in any order. However, it will be more convenient in the future if you use the convention of collecting Identification Fields at the top of the listing, followed by Information Fields.

Perhaps the most critical step in defining databases is that you need enough fields to ensure that each record in the database is unique. Redundancy in databases causes failures during access and information cross-referencing. Therefore, you need to ensure that each database has one or more fields that, combined, give unambiguous uniqueness to each record. For example, consider a sample that contains several different analytes and is assayed repeatedly for several days. If your database contained only a barcode field name, access to the record for review or editing may be confusing, if not impossible. Therefore, fields to hold the barcode, assay type, date assay performed, and batch number may be more appropriate.

You need to use your first level lists to define fields for the remainder of the information you want to keep track of and fields to use for searches, sorting, or joining databases to create relational capabilities. If your first level lists are thorough, these fields should be easy to identify.

Reviewing Field Apportionment

As you create the lists of fields in the databases for your system, it is very important to continually and thoroughly review all of the lists. When you are finished, you should review the lists together as a set. You should look for

- inconsistencies in field names.
- repeating fields within database records.
- repeating fields between database files.
- mixed relations within database records.
- redundancy in fields within database records.
- redundancy in fields between database files.
- inconsistencies in the flow of information between joined databases.
- situations that could cause records to lose their uniqueness.
- fields that are unlikely to be used.

Let us look closer at these items. Perhaps the most critical step in defining fields is that you need enough fields to ensure that each record in a database is unique. Redundancy in databases causes failures during access and information cross-referencing. Therefore, you need to ensure that each record has one or more fields that, combined, will give an unambiguous uniqueness to the record.

Reviewing database lists for repeating groups of fields and redundancy of data is also important. Repeating and redundant fields waste disk space and slow searches when indexing is not used. Repeating fields also complicate the examination and extraction of data and can lead to information being missed on searches. To illustrate the problem of ***repeating fields***, suppose the SAMPLES database of Figure 2-9 had been designed as shown in Figure 2-10. This alternate design contains two fields (ASSAY_TYP1 and

Samples Database (SAMPLES.DBF)

Record number (from @BASE)	Recno:N:8:0
Sample barcode number	BARCODE:N:8:0
Sample name/description	SAMP_NAME:C:20
Type of assay requested	TYPE_REQ:C:8
Date of request	REQ_DT:D:9
Requester's Name	REQ_NAME:C:10
Requester's phone extension	PH_EXT:N:4:0
Requester's notes and comments	RQS_NTS:C:30
Department number	DEPT_NUM:N:3:0
Project code	PROJ_CDE:C:6
Date sample will be delivered	SMP_DEL_DT:D:9
Sample rack barcode number	RACK_BCD:N:8:0
Storage location	STOR_LOC:C:10
First assay type performed	ASSAY_TYP1:C:10
Second assay type performed	ASSAY_TYP2:C:10
Date assay #1 performed	RUN_DT1:D:9
Date assay #2 performed	RUN_DT2:D:9
Batch number for assay #1	BATCH1:N:3:0
Batch number for assay #2	BATCH2:N:3:0
Plate number for assay #1	PLATE_MUM1:N:8:0
Plate number for assay #2	PLATE_MUM2:N:8:0
Dilution Factor for assay #1	DIL_FCTR1:N:10:0
Dilution Factor for assay #2	DIL_FCTR2:N:10:0
Raw data, assay type #1	RAW_DATA1:N:6:4
Raw data, assay type #2	RAW_DATA2:N:6:4
Results to be reported for #1	RESULTS1:N:10:3
Results to be reported for #2	RESULTS2:N:10:3
Sample Status	STATUS:C:10

Figure 2-10
How Not To Design A Database

ASSAY_TYP2) so that two different assay types could be performed on a sample. In this example, the ASSAY_TYP1 and ASSAY_TYP2 fields constitute a repeating group of assay type fields within the same record. Because the two fields have the same definition, the two fields are in actuality one field. If you were to scan the database to try to find all of the assays of a certain type, you would have to look in both the ASSAY_TYP1 and ASSAY_TYP2 fields because the assay type could be listed in either (or both) of the fields. Furthermore, if a new user of the system tried to get all of the records of a certain assay type, that person could easily overlook the second assay type field and miss records. Note that such retrieval problems would become even worse if you had repeating field data for more than one item.

Another problem with repeating fields brings up the question: "Where do you stop?" If you were to allow two different assay types on a sample, then why not three? If three were allowed, then why not four? If you design the system to include, for example, 10 assay types, then you would probably have many records with empty fields wasting a lot of disk space for the occasional sample that has more than one or two assay types. Conversely, you may find a sample exceeding 10 assay types. What would you do then?

For the above reasons, an alternative approach could be to have a single assay type field and create separate records in the database for each sample, each record containing identical information except for the fields dealing with the assay performed. This approach has the potential to work, but leads to a situation called a ***mixed relation***. A mixed relation contains identical data that are repeated for numerous records and can be extracted to another database. The SAMPLES database of Figure 2-10 has fields that describe background information on samples (e.g., REQ_NAME, DEPT_NUM, PROJ_CDE, and STATUS) and fields that describe assays performed on the samples (e.g., ASSAY_TYPE, RUN_DT, BATCH, PLATE_NUM, DIL_FCTR, RAW_DATA1, RAW_DATA2, and RESULTS). If you are not careful, you can create many kinds of mixed relations when blending fields aimed at different purposes in the same database. They all cause problems when you need to update your database. This statement is especially true if you are using linked files.

One such problem is the tedium of updating mixed relations. A good example of this problem is shown in the REQ_NAME and PH_EXT fields of Figure 2-10. If the requester's phone extension were to change, several thousand records would have to be updated. (Note that PH_EXT is used only as a clear illustration of mixed relations and is not intented to become part of the example database system.)

When you identify a mixed relation you should "split the relation" so that it is no longer mixed. Figure 2-9 illustrates how the database in Figure 2-10 could be split. Note that this technique eliminates both the repeating field problem and the redundancy problem that is associated with the mixed relation brought about by multiple assay types being performed on the same sample. The re-worked SAMPLES database in Figure 2-9 contains only fields that apply to background information on samples. Similarly, the RESULTS database contains only fields that apply to assay results of samples. As a consequence, the SAMPLES database now contains only one record for each sample. The RESULTS database, on the other hand, can contain several records for each sample, one record for each assay type performed on the sample. As you will see later in this chapter, the two files can then be joined using the BARCODE field. When joining the two files, the RESULTS database file needs to be the primary file and the SAMPLES database file needs to be the lookup file. ***Doing the reverse would create a situation wherein only the first occurrence of assay results for a sample would be retrievable.***

As you are looking for problems in the way Information Fields are distributed between related databases, you should take the time to ensure that the field names that will be used to join files appear in all of the files to be joined and that the names are identical in all files. For example, if databases are to be joined using ASSAY_TYPE, RUN_DT, and BATCH, the trio of fields must appear in all database files that will be joined using the trio as a basis.

Next, examine all fields for redundancy and remove fields that contain the same information. Redundant fields increase file size and make searches slower (if indexes are not used). More importantly, redundant fields invite errors during updating. That is, if several fields contain the same information, a user would have to change all of the fields during an update.

Figure 2-8 illustrates a type of redundancy that is often hard to spot. The fields containing average responses for standard curves are redundant with the curve fit parameter fields. The redundancy is due to the fact that the curve fit can be calculated from the concentrations of the standards (obtained from STANDARD.DBF through joined LOT_STD fields) and the average response fields (AVG_S0 through AVG_S7). Likewise, the statistical data could be calculated rather than archived.

When you identify fields that are redundant because they can be calculated, you often need to weigh the benefits of removing the fields. For example, performing a linear regression on a standard curve is relatively fast. However, using a nonlinear curve fitting program to find four parameters is often time-consuming. If at least one of your assays requires lengthy calculations, you should consider including fields to hold the results for all assay types. But be aware that if one of the standard responses are edited, information in the curve fit parameter fields of the database will probably be incorrect unless specifically updated.

Looking for fields that are unlikely to be used will conserve memory. If you anticipate that a field will remain empty, eliminate it from the list. Empty fields use a full allotment of memory whether they contain data or not. Empty fields also create ambiguity because the question inevitably arises as to whether the field was empty because it really should be empty, or because of neglect. A check for empty fields after a database has been in use for a period of time will often help identify fields that can be eliminated.

When looking for potentially empty fields, be careful not to remove fields that may be required for some assays, but not others. For example, the PLATE_NUM field in Figure 2-9 may be appropriate for assays performed in microtiter plates, but not for assays involving instruments with carousels. Therefore, you need to plan ahead and make fields available for all assay formats to be tracked by the system.

There is one final item to be aware of. In Chapter 3, you will be creating the spreadsheet side of a LIM system. As you create the spreadsheet, you will undoubtedly become aware of fields and/or data relationships that you have missed. Therefore, you should not consider the review you have performed in this section to be final. You should review your lists again and modify the lists based on knowledge you gain while creating the spreadsheet.

INDEX FILE SPECIFICATIONS FOR JOINING DATABASE FILES

As stated in the previous section, the problems of repeating fields and mixed relations need to be resolved. Repeating fields and mixed relations lead to redundancies that waste disk space, slow searches, necessitate excessive typing, and lead to tedious updating of databases when changes are required. The preferred method for removing repeating fields and mixed relations is to redistribute the data and create joined databases.

If you recall from Chapter 1, an extended relational database is a group of two or more files that work together in a special way. As illustrated in Figure 2-11, information

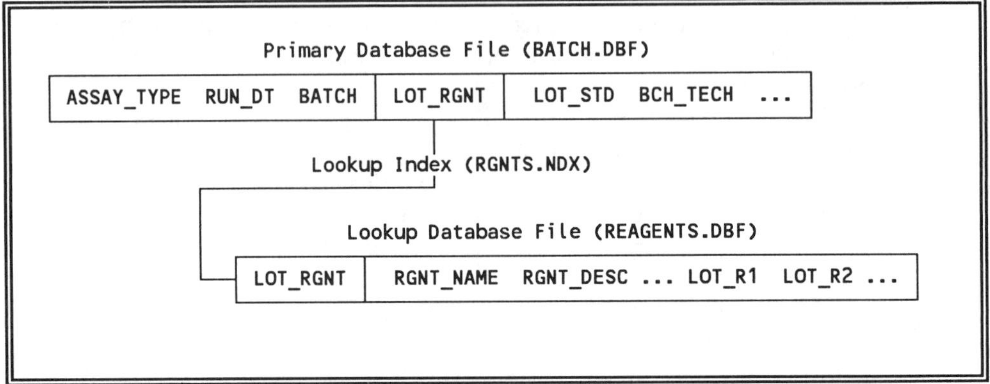

Figure 2-11

that might otherwise be stored in a single large file with many fields is instead organized in several files, each with fewer fields. The files are ***joined*** together (***linked***) so that information can be shared among the files. In a sense, one data file can look up information in one or more other files. When individual database files are linked together to extend the ready availability of related information, the resulting system appears to be a single, massive file, containing many fields. This system is often referred to as a ***virtual database*** or an ***extended relational database***.

A virtual database consists of a primary file and one or more lookup files. The information that is unique to each record is stored in a file, called the ***primary file***. Information that is shared by many primary file records is stored as a single record in a separate file, called the ***lookup file***.

To join files, there must be at least one field that is common to both files being joined (e.g., LOT_RGNT in Figure 2-11). Further, when databases are joined, the common field must have the ***exact*** same name in both the primary and lookup files. The lookup file must be indexed on the field that is common to both lookup files. The databases are joined, using the index file as an active link between the two database files. The index file link provides extended lookup capabilities.

Figure 2-11 shows a common use for a virtual database. The relational database in Figure 2-11 consists of two files. The primary file (BATCH.DBF) contains a field that is in common with the other database file (REAGENTS.DBF). The REAGENTS.DBF file in Figure 2-11 contains detailed information pertaining to the reagents used in an assay. It is common practice to prepare large quantities of reagents and use them to perform assays for several months. These reagents are given lot numbers and expiration dates for tracking purposes. Further descriptions are also pertinent: reagent name, description, preparation details, date of preparation, shelf life, lot number of components, etc. Typing all of this information into a database each time an assay is performed using the set of reagents is time-consuming, tedious, and wastes disk space. By joining the databases using the LOT_RGNT field, you can circumvent these problems.

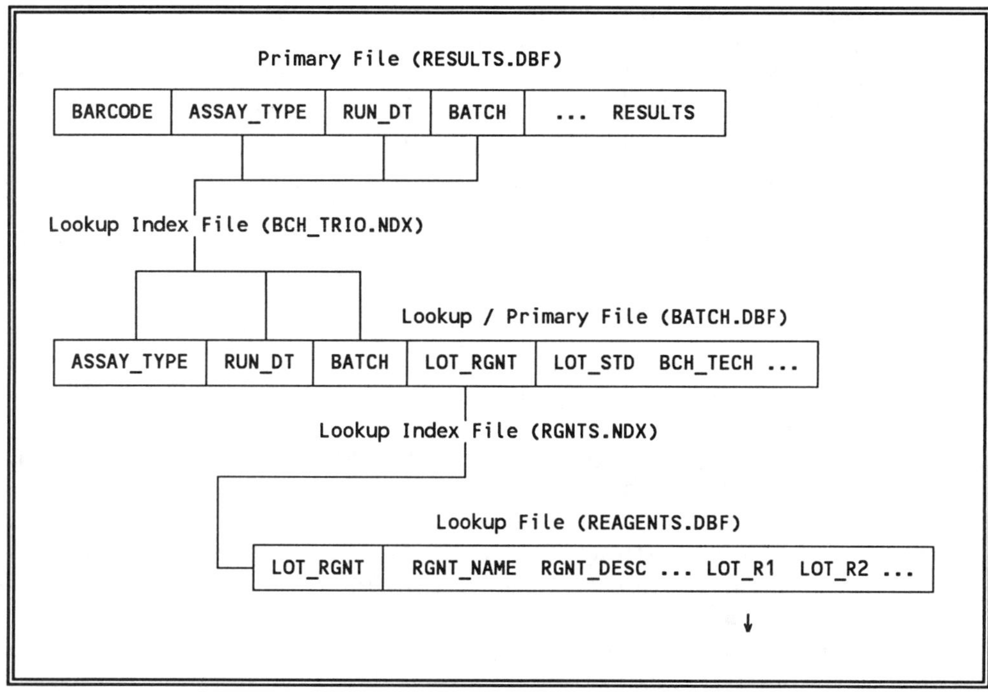

Figure 2-12

The following is a list of some important details that you need to remember when designing files to be joined:

- Two database files are joined based on an index file for the lookup file and one or more fields that are common to both files.
- Each record in the lookup file contains information that can be shared with several records in the primary file.
- The roles of primary files and lookup files should not be reversed. The primary file has many records that correspond to the unique lookup file record and only the first record is available for viewing.
- The fields selected for viewing as a result of the join (also known as ***virtual fields***) are logically added to the primary file. That is, the logically added virtual fields disappear when the session is ended.
- Virtual fields can be displayed and used in calculations, but cannot be edited.
- Field names used to join files must be identical in both files.
- Virtual field names in the lookup file must be different than field names in the primary file.

Files can be joined through several layers (as illustrated in Figure 2-12) to create a hierarchy. The hierarchy, in turn, makes field data available to each sequentially higher primary database file in the series. For example, in Figure 2-12 BATCH.DBF serves as both a lookup file (for RESULTS.DBF) and primary file (for REAGENTS.DBF). The example system illustrates how specific information about a set of reagents used during an assay run of a sample can be made available to the sample's record. The mechanism is:

- BATCH.DBF is a primary file joined to the REAGENTS.DBF lookup file using the RGNTS.NDX index file. When the files are joined, virtual fields will be selected in the REAGENTS.DBF for viewing with BATCH.DBF. (For example, R_PREP_DT and R_SHELF_LF could be obtained from REAGENTS.DBF to ensure that the reagents were not expired.)
- @BASE indexing makes selected information from REAGENTS.DBF available to BATCH.DBF.
- When viewed, a display of BATCH.DBF would contain the virtual fields from REAGENTS.DBF.
- RESULTS.DBF is a primary file joined to the BATCH.DBF lookup file using the BCH_TRIO.NDX index file. When RESULTS.DBF and BATCH.DBF are joined, virtual fields from BATCH.DBF will appear on the Select Fields screen and will be selected.
- @BASE indexing makes the selected virtual BATCH.DBF fields available to RESULTS.DBF.
- When viewed, a display of RESULTS.DBF would contain the virtual fields from REAGENTS.DBF.

Creating a hierarchy of joined files can make an enormous amount of information available to the files in a series. The higher up a primary file appears in a series, the more information it has potentially available to it. In essence, a sample record can have access to virtual fields anywhere in the LIM system as long as the system was designed and created correctly. Having extensive access to information is very desirable for LIM systems.

Advantages and Disadvantages of Joined Files

The following are advantages and disadvantages to joining files:

- *Advantages*
    - Provides an efficient means of eliminating repeating fields and splitting relations.
    - Virtual fields mean less programming needs to be performed to obtain large quantities of related information.
    - Applications run faster because several databases do not need to be searched for information. A record number in the primary database makes all virtual

field information available in its lookup index files. Therefore, separate searches are not needed and time is conserved.
  - All information pertaining to a record can be displayed together.
- ***Disadvantages***
  - More files are open, meaning more conventional memory is consumed. This consumption is especially large if your system has several joined files, large index files, index files containing multiple keys, and/or long character strings used as keys.
  - A DOS limit of 14 open files may be exceeded and system design may need to be suboptimal to decrease the total file number or to close and open files during program execution.
  - Adding records takes longer because all pertinent index files need to be updated.

Documents Required

Chapter 5 explains how to join files to implement a hierarchy of extended relational databases. Prior to initiating the concepts of Chapter 5, you will need the following documents:

- A flow diagram showing the hierarchy of joined files (Figure 2-13).
- A list specifying the content of index files to be used for joining (Figure 2-14).
- A table of virtual fields to be joined to the primary file (Figure 2-15).
- A table of search and sort keys (Figure 2-17).
- A consolidated index list (Figure 2-18).

The next subsections explain how to generate the first three documents on the list. The sections following explain how to create the last two items.

Flow Diagrams

The order in which links are created when joining files in a hierarchy is very important. The first document you will need is a flow diagram illustrating relationships of the files in the series. An example is shown in Figure 2-13. As you can see from Figure 2-13, the SAMPLES.DBF → RESULTS.DBF and BATCH.DBF → RESULTS.DBF links are higher in the flow diagram than the following links:

- REAGENTS.DBF → BATCH.DBF
- STANDARD.DBF → BATCH.DBF

Positioning affects the ability of a primary file to obtain information from lookup files in the flow diagram. As illustrated by the directionality of the arrows in the flow diagram, information from the fields in the SAMPLES and BATCH databases are available to the RESULTS database. Through the BATCH branch of the diagram, reagent and standard

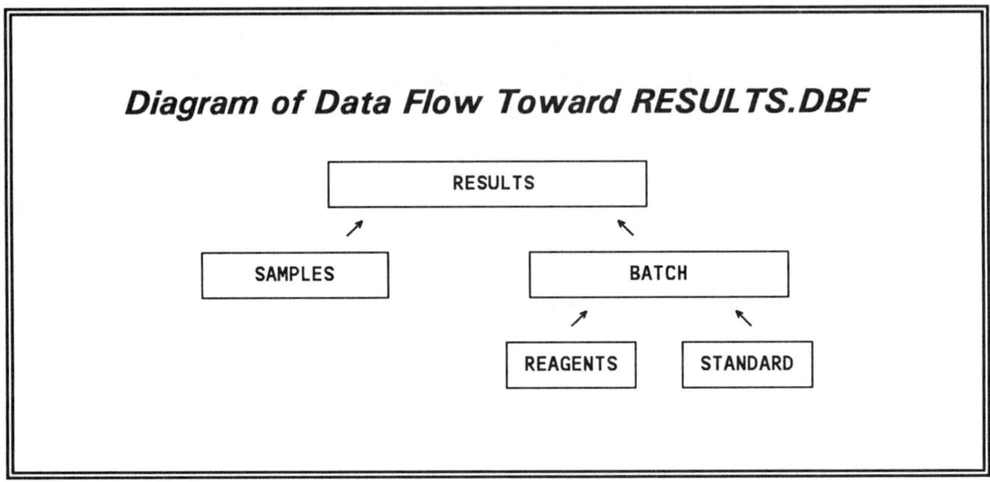

Figure 2-13

information are made available to RESULTS. The BATCH intermediate, "primary" file serves as a pipeline to make reagent and standard information available to RESULTS. This information is not available to SAMPLES. More importantly, RESULTS database information is ***not*** available to any of the database files below it in the flow diagram.

List of Data Relationships for Joining Files

The hierarchy of the flow in Figure 2-13 is reflected in the creation order for the list in Figure 2-14. The arrangement of files to be joined in Figure 2-14 is inversely related to the hierarchy of the system. To preserve the correct hierarchy, you will be starting at the top of the list in Figure 2-14 and working your way down the list when creating the links. In this case, the first link to create is the one that uses BATCH.DBF as a primary file. REAGENTS.DBF as a lookup file, and RGNTS.NDX as the index file.

Figure 2-14 also contains field specifications for joining files. If you recall, the linking process is established using an index file intermediary. In Figure 2-14, there are columns specifying

- the position of the joined databases in the hierarchy.
- the name of the primary database file.
- the name of the lookup database file containing the virtual fields.
- the name of the ***lookup*** index file to be used to mediate the join operation.
- the names of the fields in the lookup file that are to be used when creating the lookup index file.

You will need one index file for each anticipated join application. The following approach

Data Relationships for Joining Files List

HIERARCHY	PRIMARY DBF	LOOKUP DBF	LOOKUP NDX	JOIN FIELD #1	JOIN FIELD #2	JOIN FIELD #3
LOWEST	BATCH	REAGENTS	RGNTS	LOT_RGNT	-	-
	BATCH	STANDARD	STDS	LOT_STD	-	-
	RESULTS	SAMPLES	SAMPL_BC	BARCODE	-	-
HIGHEST	RESULTS	BATCH	BCH_TRIO	ASSAY_TYPE	RUN_DT	BATCH

Figure 2-14

to generating the list is one that works quite efficiently:

- Orient yourself in the correct direction. One lookup file record can service several primary file records. The lookup file contains unique information.
- Determine the primary and lookup database files you want to join from a flow diagram.
- Review the Identification Fields sections of the lookup files and determine the appropriate field names to use for the joining process.
- Check the corresponding primary files to ensure that they have the exact same field names, types, and widths for the fields to be used for joining the two databases.
- If the join field names do not appear in the Identification Fields sections of ***both*** files to be joined, see if the fields exist and if they have been assigned different names. If named differently, resolve the problem by renaming the field in one of the files. If the fields do not exist in the Identification Fields section, look for the field names in the Information Fields sections. If the field names do not exist there, add a new field to the Information Fields section to provide join capabilities.
- After all join parameters have been added to the list, re-arrange the list in reverse order of hierarchy.

To reiterate the first item on the list: It is important that you do not forget that you can only get one record from the lookup file. For example, when linking RESULTS as a primary file and BATCH as a lookup file, you can look up one batch record for many samples in the RESULTS file. However, this rule applies in both directions. Accordingly, you can look up only one sample record from the RESULTS file if the primary file is the BATCH file. Design your system accordingly.

Virtual Fields Specifications for Joining Files

PRIMARY DBF	LOOKUP DBF	FIELD #1	FIELD #2	FIELD #3	FIELD #4	FIELD #5
BATCH	REAGENTS	R_PREP_DT	R_SHELF_LF	-	-	-
BATCH	STANDARD	S_PREP_DT	S_SHELF_LF	{CONC_S0	THROUGH	CONC_S7}
RESULTS	SAMPLES	SAMP_NAME	TYPE_REQ	REQ_NAME	DEPT_NUM	PROJ_CDE
		REQ_DT	-	-	-	-
RESULTS	BATCH	{ALL FIELDS, BOTH VIRTUAL AND REAL}				

Figure 2-15

After you have created the List of Data Relationships for Joining Files, you should review the list as a whole to ensure that there are no redundant items on the list. Then, assign names to each lookup index file on the list. Try to use file names that are descriptive of the process being undertaken. Two example conventions are used in Figure 2-14. The first convention is for databases that have multiple index files. For these databases, the index file name is created by merging the lookup database file name with the first index field name and separating the two names with an underline ("_"). The second convention is usually used for databases with single index files. (Typically, databases have single index files when they are the lowest members of a hierarchy.) For this second category of databases, the index file name is created as an acronym of the lookup database file name.

Virtual Field Table

As stated previously, virtual fields

- are fields in lookup files.
- are selected for viewing as a result of a join.
- are logically added to the display of the primary database.
- can be displayed and used in calculations.
- cannot be edited.

You will need a table of virtual fields for each join that you have specified as a result of the previous section. Figure 2-15 shows a typical Virtual Fields Specifications for Joining Files list. Note that there are columns specifying

- the name of the primary database file that will display the virtual fields.
- the name of the lookup database file containing the virtual fields.
- the names of the virtual fields.

To generate the Virtual Fields Specifications for Joining Files list, review the Information Fields sections of the lookup files and determine which fields you want to view as virtual fields. Virtual field names cannot be the same as the field names in the index file being used to create the link, so you must cross-check the virtual field names against the join field names on the Data Relationships for Joining Files list (Figure 2-14) to ensure that the proposed virtual field names are not on both lists.

Next, review the field names in the primary and lookup databases. If the primary database file has field names that are identical to the virtual field names, you must change the field names in either the primary or lookup file.

INDEX KEYS FOR SEARCHING AND SORTING DATABASE FILES

Perhaps the most important aspect of the @BASE Option Pac is its support for index files. This ability is extraordinarily powerful and adds enormous value to LIM systems.

As explained in Chapter 1, each time a new record is added to a database, it is appended to the bottom of the file. If a record is to be found, the file must be searched from the beginning of the file until the record is found. This processing is time-consuming, especially for large files. If a database is to be ordered in a particular way, it must be sorted each time a new record is added. This processing is also time-consuming.

Index files can dramatically decrease sorting and searching times. This efficiency gives you substantially faster access to records in databases.

Database index files are special files that contain information about the location of records in a database. A database index is like the index for a book. A book index lets you identify the page where a topic is discussed without searching through the entire book cover to cover. A database index uses record numbers instead of page numbers, but the concept is the same. Using an index file and a ***key value***, you have quicker access to records in the database. ***A key*** is a field in an index file that points to the locations of records in a database file. The index key determines the order that the records in a data file are displayed. The index key is also used to search for records.

With an index file, adding a new record to a database involves two files. The new record is added to the end of the database as usual. In addition, the key value field and a pointer to the database record are added to the index file which is maintained in sorted order. Because two files are being updated rather than one, adding a new record to the database takes slightly more time than without an index file. However once an index file has been created, the database can be accessed very quickly because the smaller, ordered key file provides a direct pointer to the record in the larger, unordered data file.

Figure 2-16 is an example that reviews how an index file works. The index file contains a single pair of entries for each record in the database: a key value and a record number. Having a limited amount of compact data in a file makes the file much smaller

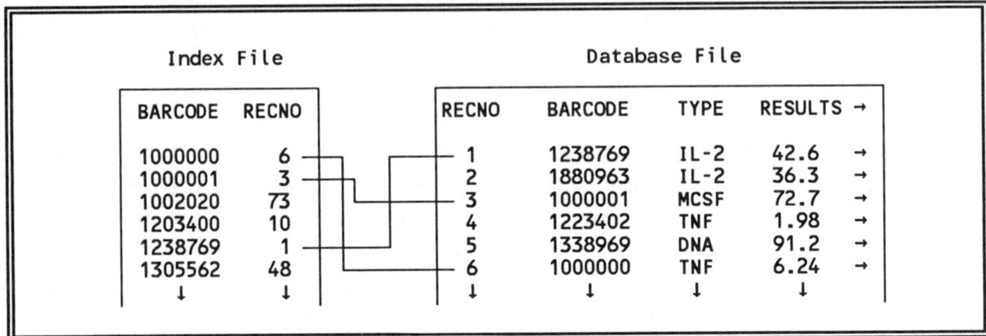

Figure 2-16

in size and substantially improves the time it takes to search the file for data. Maintaining the file in sorted order also improves search time. For these two reasons, searching through index files is much faster than searching through large, unordered database files containing immense amounts of data.

In the example system designed in this chapter, the complete record for any sample in the SAMPLES.DBF database file can be quickly located by

- first searching the index file to find the sample's BARCODE number,
- obtaining the sample's record number from the index file,
- immediately skipping to the offset in the SAMPLES.DBF database file corresponding the record number, and
- retrieving the pertinent information.

You can have more than one index per database file. It is often useful to create several index files for a single database. You can also have ***compound indexes***, where the index key consists of more than one field. For example, in addition to a BARCODE index, a second index file containing a different set of keys (e.g., STATUS and REQ_DT), might be helpful to speed preparation of end-of-month reports. If more than one index is open for a database, the primary index determines the order in which the records are found. The first index opened for a particular database is the primary index until another index is selected as the primary index.

Figure 2-17 shows a typical Search and Sort Keys list. Note that there are columns specifying

- the name of the database file to be searched and/or sorted.
- the name of the index file containing the search and sort key.
- the names of the fields to be used as keys in the index files.
- the reports that receive information from the searches and/or sorts.

Search and Sort Keys

DATABASE FILE	INDEX FILE	INDEX FIELD #1	INDEX FIELD #2	INDEX FIELD #3	REPORT
SAMPLES	SAMPL_BC	BARCODE	-	-	SAMPLE ID
RESULTS	RES_BC	BARCODE	-	-	SAMPLE RESULT
RESULTS	RES_TRIO	ASSAY_TYPE	RUN_DT	BATCH	SAMPLE-BATCH
BATCH	BCH_TRIO	ASSAY_TYPE	RUN_DT	BATCH	BATCH DETAIL
REAGENTS	RGNTS	LOT_RGNT	-	-	RGNTS DETAIL
STANDARD	STDS	LOT_STD	-	-	STDS DETAIL

Figure 2-17

You will need one index file for each anticipated search or sort application. Consider for instance, the SAMPLES.DBF database file shown in Figure 2-9. To search or sort the database on the field named BARCODE, you would need an index file specifying that field. To also perform searches using a combination of the STATUS and REQ_DT fields, you would need a second index file. If you would want to search SAMPLES.DBF using the trio of fields named ASSAY_TYPE, RUN_DT, and BATCH, then you would need a third file, and so on.

There is a three-pronged approach to generating the Search and Sort Keys list:

- Review the Identification Fields sections of each database file. Because of the nature of these sections, they are likely to contain most, if not all, of the index keys that you will need.
- Review the Search Criteria list (Figure 2-5) to ensure that all items on the list are covered by the Search and Sort Keys list. If they are not, find fields in the database listings that will provide the search and sort capabilities. If fields do not exist in the database listing, add new fields to the Information Fields sections to provide the capabilities.
- Make a review of the Information Fields listing for each database to see if there are any other search and/or sort keys that have been missed.

You should treat the Identification Fields, Search Criteria, and Information Fields lists as if they were shopping lists from which you pick items that you want to appear on the Search and Sort Keys list.

When designing your Search and Sort Keys list, try to be as thorough as possible while keeping the list as small and compact as possible. ***Remember, you can only have a total of 14 open database and index files.*** You can have more than 14 files, but that

Consolidated Index List

DATABASE FILE	INDEX FILE	INDEX FIELD #1	INDEX FIELD #2	INDEX FIELD #3
SAMPLES	SAMPL_BC	BARCODE	-	-
RESULTS	RES_BC	BARCODE	-	-
RESULTS	RES_TRIO	ASSAY_TYPE	RUN_DT	BATCH
BATCH	BCH_TRIO	ASSAY_TYPE	RUN_DT	BATCH
STANDARD	STDS	LOT_STD	-	-
REAGENTS	RGNTS	LOT_RGNT	-	-

Figure 2-18

would mean that you would have to open and close files during program execution. Opening and closing files leads to complicated and inefficient programming. This guideline is also especially important for frequently used database files. If you recall, each time a new record is added to a database file, the record is also added to the associated index files. The more index files that are associated with a database the longer it takes to maintain all of the files and the more conventional memory is consumed. If you have index files on the list comprised of search keys that are to be used only infrequently, you may want to consider removing the index files from the list and searching the database files directly.

After you have created the Search and Sort Keys list, you should review the list as a whole to ensure that there are no redundant items on the list. Then, assign names to each index file on the list. Try to use file names that are descriptive of the process being undertaken. Two example conventions are used in Figure 2-17. The first convention is for databases that have multiple index files. For these databases, the index file name is created by merging the database file name with the first index field name and separating the two names with an underline ("_"). The second convention is usually used for databases with single index files. For this second category of databases, the index file name is created as an acronym of the database file name being searched and/or sorted.

CONSOLIDATED INDEX LIST

The final step in designing your system is to merge the Data Relationships for Joining Files list with the Search and Sort Keys list to create a Consolidated Index list. An example is shown in Figure 2-18. The Consolidated Index list is a composite list that specifies index files and their contents, regardless of operation.

Because of the similar nature of search, sort, and join operations, there is usually a high degree of overlap in the Data Relationships for Joining Files and Search and Sort Keys lists. To merge the two lists, place them side by side. Review the lists for redundancy and remove all instances of it. When reviewing the lists for redundancy, proceed down the lists and compare the database file name and index field(s) of the Search and Sort Keys list with the ***primary*** database file name and the field names of the ***lookup*** index file.

That is, ***do not*** use index file names as a basis of comparison. The index file names have been assigned artificially and do not affect the process being performed. It is the database files and index field names that provide process specificity. When redundancy is found, remove the redundancy and assign a consensus name to the index file.

It should be pointed out that you do not need to finalize the Consolidate Index list at this time. The relationships contained in the list can be easily modified and without much risk at any time in the creation and use of a LIM system.

SUMMARY OF WHAT HAS BEEN ACCOMPLISHED

This chapter stressed the importance of taking time to plan your database. The approach taken was to first decide on the goals, then to decide which reports were necessary to meet those goals. From these lists, you can determine what information you need to keep track of. The required information leads to field definitions and spreadsheet layout.

For each field in a new database, the following specifications are needed:

- The name of the field.
- The field type (character, numeric, logical, or date).
- The field length (number of characters allowed).
- The number of decimal places (if field is numeric).

Perhaps the most critical step in defining fields is that you need enough fields to ensure that each record is unique. Redundancy in databases causes failures during access and information cross-referencing. Therefore, you need to ensure that each record has one or more fields that, combined, give unambiguous uniqueness to each record.

Thereafter, you need to use your lists to define fields for the remainder of the information you want to keep track of and also the fields that you want to use for searching, sorting, and joining databases.

The apportionment of Information Fields between database files is an iterative process. Field names often need to be shuffled from one file to the next before you find the right combination of pigeon holes to achieve optimal system performance and reliability.

Databases are joined in the direction in which a primary database file has many records that can look up data from a single record in a lookup file. You can join several lookup files to a primary file. By doing so, you can create very efficient database systems that have access to enormous amounts of information. For details on how to create index files for joining databases, see Chapter 5.

The following are some definitions that were provided and are important to remember:

- An ***Identification Field*** is a field used to make each record in a database unique. Identification Fields are typically used in indexes to facilitate sorting, searching, and joining databases.
- An ***Information Field*** is a field used for data archival.
- ***Repeating fields*** are two or more fields in the same database that archive essentially the same data.
- A ***mixed relation*** is one that contains identical data that pertain to numerous records and can be extracted to another database.
- A ***key*** is a field in an index file that points to the locations of records in a database file. The index key determines the order that the records in a data file are displayed. The index key is also used to search for records.
- A ***compound index*** is an index where the key consists of more than one field.
- ***Virtual databases*** are databases linked together to extend the ready availability of related information. The resulting system appears to be a single, massive file, containing many fields. Virtual databases disappear when a session is ended.
- ***Primary files*** contain information that is unique to each record. One primary file can obtain information from several lookup files.
- ***Lookup files*** contain information that is shared by many records. The shared information is stored as a single record in the lookup file.
- ***Virtual fields*** are fields selected for viewing as a result of a join. They are logically added to the primary file and disappear when a session is ended.
- A ***computed field*** is a virtual field whose value is calculated from other fields in the database. Computed fields, used correctly, can conserve disk space.

When designing a database and index system, there is a very important DOS limitation that you need to be aware of. The limitation is that although DOS allows up to 255 files to be opened, only 20 files can be accessed by each application. Because @BASE runs within Lotus, the two programs together cannot access more than 20 files, regardless of how many files are specified in your CONFIG.SYS file. Lotus and @BASE typically have about 6 open files, leaving about 14 user files. Thus, ***you can have a total of up to about 14 open database and index files before the system begins to issue error messages.***

As configured in this chapter, the current example system has five database files (REAGENTS, STANDARD, BATCH, SAMPLES, and RESULTS) and six index files (SAMPL_BC, RES_BC, RES_TRIO, BCH_TRIO, STDS, and RGNTS). It is important to allow at least one unused file so that the spreadsheet can be saved. Therefore, only $(14-5-6-1) = 2$ more open database or index files would normally be available.

WHAT'S NEXT?

Spreadsheet templates are key to the reliable and efficient flow of information between

- data acquision,
- data reduction, analysis, and graphics,
- database storage, and
- report generation.

The next chapter will show you how to make spreadsheet templates.

3

Spreadsheet Templates

Before you start creating databases, writing macro programs, or acquiring and analyzing data, you must first create a spreadsheet template to hold the data.

A spreadsheet ***template*** is simply a portion of your spreadsheet that you set aside to hold data in an organized manner. Templates usually contain only labels and formulae. Raw data can be input manually (by typing), imported from files or databases, or input directly into the cells of the spreadsheet from instruments using the concepts described in ***Laboratory Lotus: A Complete Guide To Instrument Interfacing***. Templates automatically format data and/or perform calculations. When necessary, data in templates can be transferred to database files or other templates. You can view templates in four ways:

- as "skeletons," into which your data are placed.
- as "pipelines," through which your data flow when they pass back and forth from one LIM system function to the next.
- as re-usable temporary storage facilities for data.
- as re-usable report forms that can be used in the display and/or printing of data.

You can change and experiment with data by using the templates that you create. If you have formulae in the templates, the formulae will instantly recalculate and display the results of any changes you make. This capability is extraordinarily powerful when working with experimental data and provides you with enough flexibility to be able to thoroughly examine data.

Spreadsheet templates are key to the reliable and efficient flow of information between

- user input,
- data acquision,
- data reduction, analysis, and graphics,
- database storage, and
- report generation.

If you create templates before you start programming, you will find it much easier to create programs that manipulate the data. This chapter provides you with specific design tips on how to create templates that serve multiple purposes. When you design your system, your goal is to create a single template to service as many of the five categories of data-handling just listed as possible. Typically, user interfaces and report forms need their own templates and often do not achieve the multi-purpose goal.

By designing multi-purpose templates, you will conserve memory. Multi-purpose templates also increase macro program speed and reliability because the program does not need to keep moving data from one template to another. As a general rule of thumb, the more often data are moved, the higher the chances for error.

The example templates in this chapter are very simple. Calculations need not be simple (as they are in the examples in this chapter). They can be formulae of immense complexity. Nearly all of the scientific formulae that you use can be placed into a spreadsheet template. For example, Lotus has @functions that calculate logarithms, exponentials, exponents, trigonometric functions, and routine statistics. (See Appendix A for an overview of mathematical calculations, Appendix B for an overview of Lotus @functions, and Appendix D for an overview of @BASE @functions.) You can even set up templates for matrix mathematics, polynomial linear regression analyses, integrations, differentiations, nonparametric statistical analyses, etc.

MULTI-PURPOSE TEMPLATES

Fluorescence spectroscopy is a valuable tool in almost every type of scientific laboratory, including the disciplines of biochemistry, molecular biology, enzymology, immunology, pharmacology, environmental analysis, clinical pathology, and food analysis. The example template presented in Figure 3-1 was created for an assay in which a fluorescent dye bound itself to DNA. When the dye was bound to DNA, a fluorescence appeared at 458 nanometers. A standard curve was generated using a set of standards containing known concentrations of DNA and a linear regression was performed on the curve. Unknown samples were assayed in replicate and their DNA concentrations were determined from the regression curve. This procedure was a convenient way of determining the amount of DNA in the samples.

The simple example template in Figure 3-1 can be used for acquisition, analysis, and database transfer of data for the DNA assay. Figure 3-2 shows the template of Figure 3-1 with some example data added. Veterans of previous books in this series will note a

```
-------A--------B---------C---------D---------E--------F---------G----
 1                          FLUORESCENT DNA ASSAY DATA
 2  RECORD      TYPE     RUN DATE    BATCH      TECH   STANDARDS REAGENTS
 3   Recno  ASSAY_TYPE   RUN_DT      BATCH    RUN_TECH LOT_STD   LOT_RGNT
 4
 5 TECH_COMS
 6
 7                       ****STANDARD CURVE DATA****                X
 8 NG DNA/ML  ASSAY1     ASSAY2     ASSAY3    ASSAY4    AVERAGE   PREDICT
 9         0                                                ERR       ERR
10        25                                                ERR       ERR
11        50                                                ERR       ERR
12        75                                                ERR       ERR
13       100                                                ERR       ERR
14       250                                                ERR       ERR
15
16                     ****CONCENTRATIONS OF STANDARDS****
17 Recno        CONC_0     CONC_1     CONC_2    CONC_3     CONC_4    CONC_5
18                   0         25         50        75        100       250
19
20                         ****AVERAGE FLUORESCENCE****
21 Recno       AVG_S0     AVG_S1     AVG_S2    AVG_S3     AVG_S4    AVG_S5
22                ERR        ERR        ERR       ERR        ERR       ERR
23
24              ============================================
25              |             Regression Output:           |
26              |   Constant                               |
27              |   Std Err of Y Est                       |
28              |   R Squared                              |
29              |   No. of Observations                    |
30              |   Degrees of Freedom                     |
31              |                                          |
32              |   X Coefficient(s)                       |
33              |   Std Err of Coef.                       |
34              ============================================
35
36                    *********SAMPLE DATA SECTION***********
37
38            BARCODE  DIL_FCTR RAW_DATA1 RAW_DATA2  RESULTS PLATE_NUM
39
40
41
42
```

Figure 3-1

reorganizing of the template. This reorganization places the data in a format that is still optimal for data acquisition and analysis, but expands the functionality of the template by fulfilling the requirements of @BASE for the efficient transfer of data to a database file. The reorganization of the template made it possible to fulfill the goal of designing the template to service as many of the categories of data handling as possible. In this case, all five of the categories described in the introduction of this chapter were serviced.

```
-------A--------B---------C---------D---------E--------F---------G----
1                          FLUORESCENT DNA ASSAY DATA
2  RECORD      TYPE    RUN DATE    BATCH      TECH    STANDARDS REAGENTS
3   Recno  ASSAY_TYPE  RUN_DT      BATCH    RUN_TECH LOT_STD    LOT_RGNT
4    1436 DNA         11/27/90          3 LMM        24XLM03    55D1090
5 TECH_COMS 24-HOUR INCUBATION
6
7                          ****STANDARD CURVE DATA****             X
8 NG DNA/ML   ASSAY1     ASSAY2     ASSAY3    ASSAY4    AVERAGE   PREDICT
9         0     93.9       92.9       94.3      94.3       93.9      -0.4
10       25    130.4      134.6      133.4     135.9      133.6      26.1
11       50    174.5      171.8      173.1     174.1      173.4      52.8
12       75    202.9      204.5      204.2     205.2      204.2      73.4
13      100    235.0      242.2      242.7     239.9      240.0      97.3
14      250    473.9      462.0      469.3     473.2      469.6     250.9
15
16                     ****CONCENTRATIONS OF STANDARDS****
17 Recno         CONC_0     CONC_1     CONC_2    CONC_3     CONC_4    CONC_5
18    1436            0         25         50        75        100       250
19
20                        ****AVERAGE FLUORESCENCE****
21 Recno         AVG_S0     AVG_S1     AVG_S2    AVG_S3     AVG_S4    AVG_S5
22    1436         93.9      133.6      173.4     204.2      240.0     469.6
23
24               ==============================================
25               |            Regression Output:              |
26               |   Constant                      94.518421  |
27               |   Std Err of Y Est              3.3328948  |
28               |   R Squared                     0.9994979  |
29               |   No. of Observations                   6  |
30               |   Degrees of Freedom                    4  |
31               |                                            |
32               |   X Coefficient(s)    1.494878             |
33               |   Std Err of Coef.    0.016751             |
34               ==============================================
35
36                    *********SAMPLE DATA SECTION***********
37
38           BARCODE  DIL_FCTR RAW_DATA1 RAW_DATA2  RESULTS PLATE_NUM
39           2041207        40     200.3     209.7     2956    273460
40           8268367        10     368.1     377.7     1862    273460
41           7395883         1     290.0     303.3      135    273460
42           7985840        20     427.9     428.4     4464    273460
43           1571806        10     408.9     423.2     2151    273460
44           1849327       100     106.8     113.6     1049    273460
45           7779731        40     161.8     169.1     1898    273460
46           4647388       100      94.9     102.1      266    757499
47           1250091         1     281.0     290.5      128    757499
48           7793121        20     351.8     362.4     3513    757499
49           4776316        40     250.8     269.1     4427    757499
50           5753781        40     173.5     175.7     2143    429393
```

Figure 3-2

Data acquisition and analysis usually go hand-in-hand. Both operations function similarly in terms of the flow of data into and out of spreadsheets. Because of this similarity, it is usually easy to intertwine the two functions. Not quite as easy a task is the integration of the two functions with database management. It is sometimes a little more difficult to retain an efficient functionality for data acquisition and analysis while incorporating an environment for easy database manipulation of data.

This difficulty arises from the differences in how spreadsheet-resident and database-resident data are organized. Although spreadsheets and databases are alike in that both are collections of data, they differ in the way their data are organized. Spreadsheets are basically unstructured, while databases have very rigid, well-defined structures.

The differences between spreadsheets and databases are necessary to provide a high degree of functionality to the two environments. For example, flexibility is one of the spreadsheet's greatest strengths. Any kind of data can be placed in any of the spreadsheet's cells. A well-defined spreadsheet is structured in places, but structure is not required. You can build a spreadsheet that looks like nothing more than a collection of data or calculations. Typically, a spreadsheet is well-organized, but not necessarily strictly structured. For example, blank rows are permissible in a spreadsheet to improve readability and columns of data are as acceptable as rows.

Databases, on the other hand, have very rigid structures. Each record in a row of a spreadsheet database or in a database file contains data about one item. Each spreadsheet column or database field contains the same category of data about an individual record. Typically, spreadsheet databases must be constructed such that a field name appears at the top of each column and must describe the data in that column. Blank rows and columns are not acceptable in databases. If there is a blank column between fields or a blank row between records, the database will fail. Databases contain hundreds or even thousands of records. Providing room for expansion is very important when designing spreadsheet databases.

It is essential to consider all environments when designing templates. Matrices are very useful tools that will help achieve good integrations. It is important to note the prominent role of matrices in Figure 3-1. ***Matrices are the cornerstones of good templates and are key to integrating data acquisition, data analysis, and database management.*** Assay data usually fit nicely into a matrix wherein each column of data will contain identification, data from replicate assays, results, etc., for each sample being analyzed. Each row will contain data from a single sample that has been analyzed. Data from most instruments and experiments fit into convenient matrices or formats. Data in matrices, formatted so that rows represent records and columns represent fields, promote efficient transfer of data between spreadsheets and disk-based files when using @BASE.

Because a matrix is such a major part of a template, if you make an effort to organize the matrix that holds your data now, then it will be easier to work with the data in the future. Take a moment to think about how you could organize the data from your instrument or experiment so they would fit into matrices.

Figure 3-1 illustrates efficient use of matrices to integrate data acquisition, data analysis, and database management. The next two sections provide an overview of template organization using the template in Figure 3-1 as an example. Thereafter, some specific design tips are provided. Before examining the database features in the example template, the first of the three sections looks at the data acquisition and analysis features of the template.

DATA ACQUISITION AND ANALYSIS FEATURES

The example template in Figure 3-1 is divided into five sections. These sections are typical of those found in many templates designed for applications that are used for assaying samples. The following subsections will examine the data acquisition and analysis features of each portion of the example template and ignore the database features for the moment.

The Standard Curve Section

The first data section of an assay template (after an introductory housekeeping "Header" section) is usually a Standard Curve section. This portion of the template should be designed to hold standardization data from your instrument.

In this example, the template is set up to receive data from a fluorimeter when the standard curve is being run with known amounts of DNA. It allows for quadruplicate determinations; to obtain more replicates you can add more columns. If you want fewer replicates, you could use the template as-is. No calculation will be adversely affected by fewer data points as long as the unused cells are empty and do not contain character strings. Not having strings in cells that are used for numeric data is very important. If, for example, a space (" ") has been placed in a cell, the space would be considered a string, assigned the value zero, and adversely affect the results of any calculations being performed using the cell contents.

Cells A9..A14 in the Standard Curve section (entitled "NG DNA/ML") contain concentration data for the standards. These data can

- be a permanent part of the template.
- come from user input.
- be transferred into the template from other areas of the spreadsheet by macro programs.
- be extracted from a database using @BASE @DB and @NDX functions nested within a macro program.

Cells F9..F14 in the Standard Curve section (entitled "AVERAGE") contain active Lotus @AVG @functions that provide averages of the raw data points. A good alternative to using active formulae is to use macro programs to perform the calculations and place the results in template cells (see Chapter 6). Macro programs are usually better than active formulae because spreadsheets recalculate faster and utilize less memory when numbers are in cells than when formulae are in cells.

The column entitled "X PREDICT" gives the concentration values that the known amounts would have given if they were calculated from the regression curve. The "X PREDICT" column provides a good visual check for how well the regression curve "fits" the data. That is, this column allows you to compare the values predicted from a regression curve to actual values. The closer the actual values agree with the predicted values, the more accurate your regression analysis is, and the more accurate the predictions for your samples will be.

Cells B18..G18 and B22..G22 of the Standard Curve section contain information that is redundant with cells A9..A14 and F9..F14, respectively. This redundancy makes it appear as if both sets of tables would not be needed. In many situations, that case is precisely correct. However, in other situations, it is convenient to supply both column and row tables of the same data in a template (whether the second table is part of the template report or not). The tables in this example have been included to illustrate one such situation and will be re-examined in the Database Features section to illustrate some other important points.

The situation necessitating the redundant tables in this example arises from differences in data formats required by the Lotus regression and @BASE data transfer programs. If data are to be used by the Lotus regression program, the data must be in columns, not rows. This requirement is at odds with the @BASE menu-based transfer of data, which requires that data be in rows, not columns. Therefore, a compromise must be made. Either the transfer must be made using @DB and @NDX @functions from a column, or the template needs to contain the data in both formats. Having both formats in the same template is redundant, but allows regression analysis and speeds database transfer.

Cells B18..G18 of the Standard Curve section (entitled "CONCENTRATION OF STANDARDS") contain concentration data for the standards. As was the case for cells A9..A14, these data can come from a variety of sources. Further, the cells in row 18 can be a source for data transfer to cells A9..A14. To copy data from cells B18..G18 to cells A9..A14 and simultaneously re-orient the data from row to column format, you could use the /***R***ange ***T***ranspose sequence of commands.

The table containing cells B22..G22 of the Standard Curve section (entitled "AVERAGE FLUORESCENCE") is in a correct format for @BASE to efficiently transfer data to a database using menu-based commands, thereby providing a convenient pipeline for data transfer. The data in this table can originate from any of the sources described previously in this section.

If redundant tables in a template cause confusion, they can be hidden using a /***R***ange ***F***ormat ***H***idden (Lotus 1-2-3) or /***R***ange ***O***ther ***H***idden (Symphony) command sequence. Alternatively, the tables can be placed elsewhere in the spreadsheet, outside of the viewing area.

The Curve Fitting Section

The next section of an assay template is usually a curve fitting section for the standard curve data. In this example, the parameters for the standard curve are determined using a linear regression. Both Lotus spreadsheets have the ability to perform a broad range of regression analyses on data. The Lotus regression program is very powerful and will support almost any type of regression that you would need to perform. For example, Lotus can do exponential, polynomial, and log regression analyses. For other forms of data, powerful macro programs can be developed to perform nonlinear curve fitting, interpolation, etc. (See ***Practical Spreadsheet Statistics and Curve Fitting for Scientists and Engineers*** for more details on the Lotus regression program and macro programs that can be used for curves that are not linear in their coefficients.)

The Regression section of this example consists of an area in the template that receives data from the Lotus regression program. However, when the numbers for the regression parameters are returned from the regression program, the program places text into the spreadsheet along with the numbers. The text appearing in the box of Figure 3-1 originated from the regression program.

If you were to enter your own text in this area, it would be overwritten. You can circumvent this problem by building a box in the format that you want, picking an area of the spreadsheet outside the viewing area, directing the regression program to place the regression numbers into that area, and placing formulae into the cells in the template that refer to the corresponding cells in the regression area.

For example, if your regression area had cell Z25 as its "constant" cell, you would place +Z25 in the cell of your template that corresponds to this piece of data.

If you recall, the last column of the Standard Curve section (the one entitled "X PREDICT") relates to the Regression section. Later, when you create the template for this example, you will place Lotus equivalents to the formula $x = (y - b)/m$ into the cells of this column. (When you create the template for your application, use equations that appropriately describe the data for ***your*** standard curve.)

When a linear regression is performed on the "NG DNA" and "AVERAGE" columns in this example, the "X PREDICT" column updates using the "Constant" (intercept) and "X Coefficient" (slope) from the linear regression. Using the results, you can see how closely each of the points on your standard curve agree with those predicted by the regression analysis.

The Raw Data and Sample Data Sections

The final two sections of an assay template are the Raw Data and Sample Data sections. In the example template, these two sections have been combined to facilitate transfer of data to a database (see below).

The Raw Data portion of this composite section is set up to receive fluorescent units data for the samples. In the example, space is included for duplicate determinations on samples containing unknown amounts of DNA. This section may be different for other experiments and/or applications. The beauty of a spreadsheet is that you can create custom templates to fulfill most purposes.

The Sample Data (Results) portion of this section contains cells for DNA concentrations calculated from the linear regression curve. As was the case in the Standards section, the formulae in this section use the X Coefficient and Constant from the regression analysis to determine the DNA concentration for a sample. The formulae are a re-arrangement of the familiar $y = mx+b$ formula and are the Lotus equivalent to $y = (y - b)/m$. Again, this formula may change for your particular situation.

DATABASE FEATURES

As stated above, templates are convenient pipelines through which to pass data

- from data acquisition and analysis to database files.
- from user input to database files.
- from database files to display and hard copy reports generated as results of searches (queries) and sorts.

This section addresses the first type of data pipeline (from data acquisition and analysis to databases). In later sections, the two other pipeline types will be described. However, the basic concepts are the same for all forms of data transfer.

With @BASE, there are two basic methods to make transfers between spreadsheets and database files:

- Use the @DB and @NDX @functions within macro programs to transfer data back and forth. With this method, specifically designated cells can be serviced anywhere in the spreadsheet. This method is not very efficient because relatively slow Lotus {PUT} commands and @index @functions need to be used. However, the method is acceptably fast when transferring small amounts of data and provides a high degree of flexibility when needed.
- Use the transfer utilities of the @BASE menu (***D***ata ***T***ransfer ***E***xport, ***D***ata ***T***ransfer ***I***mport, ***O***ptions ***I***ndex ***D***ata ***E***xport, ***O***ptions ***I***ndex ***D***ata ***I***mport, and ***O***ptions ***I***ndex ***D***ata ***K***ey-Import). These utilities are very efficient, but the template data must be strictly structured before they can be used.

When using the menu-based transfer utilities of @BASE, the template used for transfer must be structured so that it conforms to the rigid rules imposed by @BASE. The template must be a two-dimensional table with a two-part structure, consisting of

- a header of field names, followed immediately by
- a body of data.

The table need not contain all of the field names specified in the database, but there must be no repeating field names. Further, there can be no blank rows or columns within the table. Figure 3-1 illustrates implementations of this format in the Header, Standard Curve, and Sample Data sections. The format is discussed further in subsequent subsections.

Data can be imported and exported by

- using record numbers (Recno).
- using key values (values corresponding to the fields in the index being used).

If Recno is present in the field name header, the following @BASE menu utilities will use record numbers to transfer data:

- *D*ata *T*ransfer *E*xport
- *D*ata *T*ransfer *I*mport
- *O*ptions *I*ndex *D*ata *K*ey-Import

Recno is not a valid field name for the following @BASE menu utilities:

- *O*ptions *I*ndex *D*ata *E*xport
- *O*ptions *I*ndex *D*ata *I*mport

These two utilities can only use key values as a basis for data transfer.

The choice of whether to use Recno or key values depends on what you are trying to achieve during the transfer process. Following is some information that will help you decide whether to include Recno in field headers being used for transfer:

- When exporting data using Recno as a basis, @BASE looks at the Recno field for each record in the table being exported and compares it against the database. If there is already a record with that number in the database, the record is replaced with the one from the spreadsheet. If the record number does not match, a new record is added to the end of the database file using the data from the spreadsheet. Therefore, Recno can be used to change existing records and add new records to a database in one operation.
- Record number-based export is commonly used in conjunction with record number based import. Record number based import is used to transfer records to a spreadsheet so that they can be modified and transferred back to a database file with record number-based export utilities. Therefore, if you plan to import and modify records in the spreadsheet and then export the records back to a database file, it is important to include Recno fields in the template.
- When transferring data from a database to a spreadsheet using @DB @functions, records are specified using the Recno field. Therefore, having Recno information available at appropriate locations in a template is also important for @DB @function data transfer.
- Record number-based export can be used to target specific records when multiple records have the same key value. If a record number was determined and placed in a column headed by "Recno," exporting utilities would overwrite the existing data of the target record. Thus, if your program found record numbers at the beginning of a program, the program could be used to overwrite existing data.

- Using record numbers can also be used to create several records with the same key value. A common example would be one in which several different assay types were performed on the same sample. If the index key-based utilities were used, exporting data would overwrite the record with the first occurrence of the key, but new records would be created if the record number utilities were used and no Recno column was used or if a zero was placed in the Recno column.
- Using key values is the best choice if records with unique key values are present in a database and you want to update (overwrite) existing records.
- Using key values is the best choice if you want to keep records unique (i.e., only one record per key value).

To use key values, the first entries in the field label row must contain the field names used in the index file. (Again, you ***CANNOT*** use the Recno field for key value-based transfer utilities).

When data are exported to, or imported from, a database file using record numbers as a basis, the first field name must be "Recno". Preceding data to be exported with Recno causes @BASE to look at the Recno field for each record in the spreadsheet, compare it against the database, and apply the following rules and consequences:

- If there is already a record with that number in the database, the existing record is overwritten and updated with the data in the spreadsheet.
- If the record number does not match one in the database file, a new record is added before the data from the spreadsheet are transferred.
- If a Recno field and associated record numbers do not preceed the data, new records will be created.

Similarly, if the first field name is Recno, record number-based importing utilities will automatically transfer record numbers to the table.

The next subsections discuss details of some of the other database features of Figure 3-1. The goals of the features are to facilitate the update of all fields in the BATCH.DBF and RESULTS.DBF databases defined in Chapter 2. The field definitions for these two files appeared in Figures 2-9 and 2-12, respectively. For your convenience, the field definitions have been reproduced in Figures 3-3 and 3-4, respectively.

Header Section

The Header section serves two purposes:

- It displays relevant information about the batch.
- It serves as a temporary storage facility for information necessary for @BASE data transfer between the spreadsheet and database files.

Let us discuss the second purpose in more detail. Note the structure of the table in the

Assay Batch Database (BATCH.DBF)

Identification Fields

Record number (from @BASE)	Recno:N:8:0
Assay type	ASSAY_TYPE:C:10
Date assay performed	RUN_DT:N:5:0
Batch number	BATCH:N:3:0
Lot number of standards used	LOT_STD:C:20
Lot number of reagents used	LOT_RGNT:C:20

Information Fields

Technician performing assay	BCH_TECH:C:5
Average response for standard #0	AVG_S0:N:8:4
Average response for standard #1	AVG_S1:N:8:4
Average response for standard #2	AVG_S2:N:8:4
Average response for standard #3	AVG_S3:N:8:4
Average response for standard #4	AVG_S4:N:8:4
Average response for standard #5	AVG_S5:N:8:4
Average response for standard #6	AVG_S6:N:8:4
Average response for standard #7	AVG_S7:N:8:4
Curve fit-A Parameter/Slope	CURV_A:N:8:4
Curve fit-B Parameter/Intercept	CURV_B:N:8:4
Curve fit-C Parameter	CURV_C:N:8:4
Curve fit-D Parameter	CURV_D:N:8:4
Standard Error of Y Estimate	STD_Y:N:8:4
Correlation Coefficient Squared	R_SQR:N:6:4
Number of Observations	N_OBS:N:3:0
Degrees of Freedom	DF:N:3:0
Standard Error of Coefficient	STD_C:N:7:4
Average response for control #1	AVG_C1:N:8:4
Average response for control #2	AVG_C2:N:8:4
Date of final review by supervisor	FREV_DTE:D:9
Technician comments	TECH_COMS:C:21

Figure 3-3

Header (cells A3..G4). It conforms to the rules imposed by @BASE for Recno data import and export. The table consists of a row of field names, with the first field name being "Recno", and is followed immediately by a body of data. Further, there are no blank rows or columns within the table.

As stated above, @BASE Recno data transfer to database files depends on Recno data. The Header table contains data that facilitate a search of an existing database to ascertain whether a record exists for the ASSAY_TYPE, RUN_DT, and BATCH combination. If you recall from Chapter 2, this trio of fields fully characterizes a batch and appears in the BATCH.DBF database file. The trio in the Header table can be used to verify whether a record exists for the batch and, if not, create it. Regardless of origin, the record number data can then be imported into the Header section and later transferred to other sections of the template. Recno can then be used in updating other fields of the database.

Assay Results Database (RESULTS.DBF)

Identification Fields

Record number (from @BASE)	Recno:N:8:0
Sample barcode number	BARCODE:N:8:0
Type of assay performed	ASSAY_TYPE:C:10
Date assay performed	RUN_DT:N:5:0
Batch number	BATCH:N:3:0

Information Fields

Plate number	PLATE_NUM:N:8:0
Dilution Factor	DIL_FCTR:N:10:0
Raw data point #1	RAW_DATA1:N:8:4
Raw data point #2	RAW_DATA2:N:8:4
Results to be reported	RESULTS:N:10:3
Results Status	R_STATUS:C:4

Figure 3-4

To illustrate this use, the data in the Standard Curve and Curve Fitting sections of the template will be used to update records in the BATCH.DBF file. When a macro program is used with the template, the program copies the record number from the header section to the CONCENTRATIONS OF STANDARDS and AVERAGE FLUORESCENCE sections. Then, when data are transferred to the BATCH.DBF database file using a Recno-based transfer utility, @BASE looks at the Recno field transferred into the tables and compares it against the database. If there is already a record with that number in the database, the existing record is overwritten and updated with the data in the spreadsheet. If the record does not match, a new record is added before the data from the spreadsheet are transferred.

Similarly, @DB @functions are used to update BATCH.DBF with curve fitting data. These @functions require record numbers to specify the records to update. This information is made available in the Header section.

If you also recall from Chapter 2, the trio of fields also appears in the RESULTS.DBF database and provides a link between individual sample data and the BATCH.DBF file. By extending the Sample Data section to the right of the report form (columns H through J), including the field names of the trio in cells H38..J38, and copying the cells in the Header section to each record in the SAMPLE DATA section, the information can be added to each sample's record when @BASE performs updates of RESULTS.DBF.

The remaining fields in the Header section contain the initials of the technician performing the test and lot numbers for the standards and reagents, respectively. This information is placed in the table for easy transfer to the BATCH.DBF file defined in Figure 3-3.

The Standard Curve Section

As noted previously in the Data Acquisition and Analysis Features section, the two tables at the bottom of the Standard Curve section contain redundant information. Nonetheless, the tables were included because they are in correct format for efficient @BASE menu-based data transfer between spreadsheets and databases.

The two tables in the Standard Curve section interact with databases designed in Chapter 2. The CONCENTRATION OF STANDARDS table interacts with the STANDARD.DBF database detailed in Figure 2-11, using LOT_STD in the Header section to locate the appropriate record. The field names at the top of the table (e.g., CONC_S0, CONC_S1, ..., CONC_S5) correspond to field names in Figure 2-11. As will be shown later in this book, field names in this format can be used in macro programs to facilitate specification of data fields to be imported. The data in the row below the field names are updated using a macro program and @BASE file transfer utilities to import the necessary information.

The AVERAGE FLUORESCENCE table formats data to correspond to definitions appearing in BATCH.DBF. The field names at the top of the table (e.g., AVG_S0, AVG_S1, ..., AVG_S5) correspond to field names in Figure 3-3. Immediately below the field names is an open area set aside for the data to be transferred. This format and the leading Recno field facilitate data export to BATCH.DBF.

The Curve Fitting Section

The Curve Fitting section also contains data required by BATCH.DBF. Because the Lotus regression program places data in the spreadsheet using a pre-defined format, menu-based export to BATCH.DBF is impossible without substantial re-formatting. For this reason, the regression data are used as-is and @DB @functions are normally used for database transfer. If custom curve fitting macro programs are used to place data in Curve Fitting sections, the programs should do so in the tabular format imposed by @BASE.

The Recno from the Header section is required to specify the record to be updated in BATCH.DBF. The data fields in BATCH.DBF extracted from the Curve Fitting section are:

- CURV_A (X Coefficient; slope)
- CURV_B (Constant; intercept)
- STD_Y (Std Err of Y Est)
- R_SQR (Square root of R Squared)
- N_OBS (Number of Observations)
- DF (Degrees of Freedom)
- STD_C (Standard Error of Coefficient)

Note that the combination of data in the Header, Standard Curve, and Curve Fitting sections contain all of the information needed to fully update the BATCH.DBF database file. In the next subsection, the data template servicing the RESULTS.DBF database file will be examined.

The Raw Data and Sample Data Sections

The data in the Sample Data section are formatted for easy @BASE menu export to the RESULTS.DBF database file. Careful formatting of data in sections of this nature is essential because substantial numbers of records are normally exported from the sections to database files. Therefore, a high degree of efficiency is needed for the transfer. The use of @DB @functions would be unacceptably slow because they would require the use of inefficient Lotus @INDEX @functions. To achieve efficiency, @BASE menu-based export of data is used. To use menu-based export, the table must conform to the rules set forth by @BASE.

The SAMPLE DATA table formats data to correspond to definitions appearing in RESULTS.DBF. The field names at the top of the table (e.g., BARCODE, DIL_FCTR, RAW_DATA1, RAW_DATA2, RESULTS, PLATE_NUM) correspond to field names in Figure 3-4. Immediately below the field names are the data to be transferred. The data table contains no unfilled rows. This format facilitates data export to RESULTS.DBF.

Note the important absence of a Recno field name at the top of the table. In this case, new records would always be created when record number-based utilities were used. Forcing creation of new records is a common prerequisite for sample data in which several different assays may be performed on each sample. Another situation is when assays will be performed on the same samples on more than one day.

As noted at the beginning of this section, there are fields to the right of the template shown in Figure 3-1. The fields contain data copied from the Header section and include ASSAY_TYPE, RUN_DT, and BATCH. If you recall from Chapter 2, this trio of fields appears in the RESULTS.DBF database and BCH_TRIO.NDX index files and provides a link between individual sample data and the BATCH.DBF file. (See Figures 3-4, 2-14, and 2-7.) By extending the Sample Data section to the right of the report form, including the field names of the trio, and copying the cells in the Header section to each record, the information will be added to each sample's record when @BASE performs updates.

Extra fields have been placed to the right of the template so that the template can be used as a report form. By doing so, the report form can be viewed and printed without including a large block of repetitive data that is also redundant with information appearing in the Header section. This arrangement makes the report form more pleasing to the eye and less confusing. Another technique that can be used to seclude data is to place the data in columns and use a Lotus menu command to hide the column. For Lotus 1-2-3, the command sequence is /***W***orksheet ***C***olumn ***H***ide. For Symphony, the command sequence is /***W***idth ***H***ide. Lotus also allows ranges of cells to be hidden. Hiding ranges of cells is particularly useful when data are surrounded by other data that need to be displayed. For Lotus 1-2-3, the command sequence is /***R***ange ***F***ormat ***H***idden. For Symphony, the command sequence is /***F***ormat ***O***ther ***H***idden.

Another important feature of the Sample Data section is the placement of the sample data matrix. Typically, sample data matrices are at the bottom of templates. This organization allows the matrices to be open-ended so that they can be expanded for variable numbers of samples. Placement of a sample data matrix at the bottom of a template is especially important if hundreds, or even thousands, of samples are anticipated. A situation that creates large numbers of samples and is often overlooked is one in which a template serves the alternative purpose of holding summary data retrieved from a database. When summary data are retrieved (e.g., all samples assayed between two dates), many thousands of records may be placed in the template. If the Sample Data section was not at the bottom of the template, other data would be overwritten.

The next section reviews and expands on the style considerations provided in the previous discussion of the example template. The tips should be considered when designing your templates.

PLANNING YOUR TEMPLATES

To make the best use of a template's power, take the time to ***plan***. Plan your templates using the concepts illustrated in the previous sections. It is of utmost importance to try to design templates that achieve the greatest harmony between data acquisition, data analysis, and database management. Whenever possible, try to create templates that promote efficient transfer between spreadsheets and database files when using @BASE.

The time that you spend planning your templates will be more than offset by the time that you will save when you begin programming. If you do not properly plan your templates and how they fit into your spreadsheets, you may have to make changes that require inserting rows, deleting columns, moving data to different cells, etc. Re-designing templates and rewriting programs take a lot of time and are prone to errors.

When you design your templates, review the lists prepared in Chapter 2 to ascertain all data that need to be in the templates. Then, determine the location of each piece of data. Pay particular attention to whether the data would be better presented horizontally across the spreadsheet or vertically down. From this decision, you will be able to determine the locations for any macro programs that you plan to use. Coordinating the positions of the two is important because the template and its interactive program must form an efficient unit that exchanges data with the greatest harmony and safety. Remember always that a well-planned template is the key to efficient programs.

You will also want to organize the areas of the template so that programs can efficiently place data into them. For example, you may want a section for standards responses, a section for curve fitting, a section for unknown sample responses, and a section for final results.

In addition, remember that experiments tend to increase in size as projects proceed and new ideas are generated. If you plan your template for expansion from the start, you can reserve room on the spreadsheet for extra rows and columns that can be used to hold extra data.

The following is a list of some review and supplementary style guidelines to consider when designing templates:

- Always design template sections so that they are formatted with efficient database transfer in mind. Data transfer between spreadsheets and database files is usually the slowest process in a LIM system. Unfortunately, data transfer is also one of the functions that is most often performed.

- When designing templates, do so while looking at the lists and data flow diagrams created in Chapter 2 to ensure that all appropriate fields of data are included in the template.
- Whenever possible, format data in tables that conform to the rules imposed by @BASE for data menu transfer. Data menu transfer is faster and often easier than @DB and @NDX @function transfer because of the requirement to use inefficient Lotus {PUT} commands and @INDEX @functions.
- A table used for @BASE data menu transfer should consist of a header of field names (with the first field name being "Recno" if transfer is record number-based) followed immediately by a body of data. Further, there must be no blank rows or columns within the table.
- When using field names across the top of a table for @BASE data menu transfer, all field names need not appear in the table or even be in the same order as the database, but all field names ***must*** be spelled identically to those in the database. Because field names tend to be acronyms and therefore somewhat confusing, it is often desirable to hide the field names and use more descriptive names in the row immediately above the field name row. For Lotus 1-2-3, the command sequence for hiding a range is /***R***ange ***F***ormat ***H***idden. For Symphony, the command sequence is /***F***ormat ***O***ther ***H***idden. (This treatment has not be applied to the example templates in this book so that you can see the field names. However, under normal circumstances the range names would be hidden.)
- If virtual field names are used, the names must be in lower-case letters.
- In templates that are involved in record number-based data transfer, do not forget to include fields in the spreadsheet for a Recno.
- The more formulae that you have in cells, the slower the spreadsheet will recalculate. If you have several hundred formulae in a template, every time the spreadsheet recalculates it will take a substantial amount of time. Therefore, if you need to place large data summaries in a spreadsheet, you should have a program do as many of the calculations as is practical.
- It is especially important to consider the above guideline when using @BASE @DB @functions. Each time Lotus performs a spreadsheet recalculation, @BASE recalculates all @DB @functions in the spreadsheet. Every time an @DB @function is evaluated, @BASE starts at the beginning of a database file and reads each record until it locates the information it is searching for. This search is typically a very slow process. For example, it would take approximately 9 seconds to evaluate a search on 1,000 records in a database on a standard IBM AT running at 6 Mz. If a Lotus spreadsheet contained many @DB @functions, and if a database file contained thousands of records, the spreadsheet would take a very long time to recalculate. ***Therefore, you should avoid at all costs the use of @DB @functions in spreadsheet cells.*** Instead,

"nest" the functions in {LET} macro commands. By doing so, the @BASE @functions will be calculated only when macro commands execute and spreadsheet recalculation will not be affected.

- @BASE @NDX @functions are much faster than @DB functions. For reasons explained in Chapter 2, @NDX @functions are nearly instantaneous. Therefore, @NDX use in cells would not significantly slow spreadsheet recalculation. However, @NDX @functions are more commonly nested in {LET} commands of macro programs, rather than used directly in cells of spreadsheets.
- Take the user's ability into account. Use headings and labels that are easily understood. Use as little jargon as feasible. If you use non-descriptive terms, it may confuse your audience. This confusion will lead to inefficient usage and eventual errors that will waste time and cause serious errors in the data (and therefore, the interpretation of the experiment).
- Do not place too many different types of experiments in one spreadsheet. Crowded layouts can make spreadsheets cumbersome and confusing for the user. Try to keep templates as small and uncluttered as possible.
- Sometimes a little planning will permit you to design a template that will serve multiple purposes. For example, you may be able to get day-to-day performance data for quality control purposes from the same template that holds your experimental data. Or, if you work in a centralized assay laboratory that performs sample analyses from several other laboratories, your need to issue individualized reports (that contain only the data that a submitting lab is interested in) can be met by laying out a template that can be individualized. (You can use the /*W*orksheet *C*olumn ***H***ide command (Lotus 1-2-3) or the /*W*idth ***H***ide command (Symphony) to hide data from other submitting laboratories and just print the appropriate data for the lab in question. Lotus also allows ranges to be hidden. Hiding ranges is particularly useful when data are surrounded by other data that need to be displayed. For Lotus 1-2-3, the command sequence is /***R***ange ***F***ormat ***H***idden. For Symphony, the command sequence is /***F***ormat ***O***ther ***H***idden.)
- If long lists of records need identical information in a table used for exporting data, you should first ask whether the data could better be contained in a joined database. If not, then the data should be copied to columns immediately to the right of the existing table. The data could then be concealed by hiding the columns.
- If data are to be used by the Lotus regression program, the data must be arranged in columns, not rows. This arrangement is at odds with @BASE menu-based data transfer, because data must be in rows not columns for this type of transfer. Therefore, a compromise must be made. Either the transfer must be made using @DB and @NDX @functions from a column, or the template needs to contain the data in both formats. Having both formats is redundant, but makes database transfer faster.
- If a date is needed in a template, the date can be obtained from the Lotus @NOW @function. If the @function is placed in a cell of the template, it will

update continually. If the @function is placed in a macro program, it will update only when the macro program executes. Therefore, you must decide how you want to use the date. If the current date is all that is needed, then it can be placed in a template cell. If the date is to be used for database update, it should be placed in a macro program.

- If a cell will be displaying a date, the cell must have a width of at least 9 characters and formatted correctly. To format dates in Lotus 1-2-3, use the /***R***ange ***F***ormat ***D***ate ***4***(Long Int'l) command. To format dates in Symphony, use the /***F***ormat ***D***ate ***4***(Full Int'l) command.

After you have designed your templates, it is very important to thoroughly review all of the field name lists and flow diagrams created as a result of Chapter 2. Review them to ensure that all data needed to update the relevant databases are contained in the templates. You should also review the lists for correctness and completeness. It is often the case that the creation of templates exposes errors in the lists. In fact, generating templates is a valuable tool for ensuring the appropriateness of field name lists.

SIGNIFICANT FIGURE AND ROUNDING CONSIDERATIONS

You may want to include significant figure considerations in your templates. That is, if the data that come from your instrument or experiment are good only to a certain number of significant figures, then you would probably not want to display data past that number of digits.

Most laboratories have significant figure and rounding guidelines that they follow for data reduction. However, depending on the discipline, the level of accuracy/precision required, and the level of significance that the rules play in the accuracy of the data to be reviewed, there is great variability in the rules employed. Because of the importance of significant figures and rounding in obtaining accurate values in data reduction, you may want to review the significant figure discussion in Chapter 3 of ***Practical Spreadsheet Statistics and Curve Fitting for Scientists and Engineers***.

To summarize the discussion, if you want full precision in your calculations, but prefer to display fewer decimal places, use the /***R***ange ***F***ormat ***F***ixed or the /***W***orksheet ***G***lobal ***F***ormat ***F***ixed command sequence (Lotus 1-2-3). For Symphony, use the /***F***ormat ***F***ixed or the /***S***ettings ***F***ormat ***F***ixed command sequence. Lotus will ***display*** only the number of decimals specified, but will ***preserve*** all digits during calculations. This method can cause errors, especially if tests of equality are being performed on data. If you need to strictly adhere to significant figure rules during calculations, use @ROUND @functions or prepare special formulae (based on @ROUND). Rounding is especially important if you will be testing data for equality. If you need only the integer portion of numbers and want the numbers truncated, use the @INT @function.

OVERVIEW OF CREATING TEMPLATES

Creating any template is easy after you have planned it. You just type in text headings and formulae, create named ranges, and specify formats for specific cells.

Because a template is nothing more than a completed "skeleton" without data, you must test a template by manually entering characteristic data into it to make sure that it works correctly. Then, after all of the labels, formulae, and programs have been entered and the template is working correctly, store it. ***Store the spreadsheet template without data.*** With a "clean slate" template, you will not have to worry about deleting entries for each subsequent use.

You should also create copies of the spreadsheet on several floppy disks and store them in separate locations. These backup copies will be a safeguard against having to re-create the spreadsheet if the current one is damaged or lost.

CREATING THE TEMPLATE FOR THIS EXAMPLE

The following instructions will teach you how to create the template shown in Figure 3-1. The steps you will use to create your template will be essentially the same. Creating this template will teach you some important concepts that you can apply when you create ***your*** template.

Column titles are centered. The spreadsheet will automatically perform the centering task if you issue the following commands. In Symphony, issue the /*S*ettings *L*abel-Prefix *C*enter command sequence. In Lotus 1-2-3, the command sequence is /*W*orksheet *G*lobal *L*abel-Prefix *C*enter.

Using the arrow keys, move the cell-pointer to the appropriate cells and type in the relevant text or number. Ignore the cells containing "ERR" for the moment.

Use numerals instead of labels whenever you create column or row headings that will be used in calculations or graphics to format the headings in an appropriate form. If labels are used, Lotus treats them all as character strings, assigns their values to zero and does not give proper calculations or graphics displays.

To create the box, use a single quote and a series of equal signs for the horizontal lines ('===). Use a single quote and shift-backslash ('|) for the vertical lines. For now, leave the box empty. Later, when the template is tested, text will automatically be filled in by the Lotus regression program. More instructions are listed below:

1. Into cell F9, type the formula to calculate the average Fluorescent Units value in Row 9:

```
@AVG(B9..E9)
```

   Issue the /*C*opy command to copy the formula in cell F9 to Rows 10 through 14. When you are prompted for the cells to copy from, type F9 and press [RETURN]. When prompted for the cells to copy to, use the arrow keys to move to cell F10, type a period (.), and use the down arrow key to move to cell

F14. Press [RETURN].

ERR will appear in all of the cells because the ranges on which their calculations depend do not have values yet. This display will appear in cells of your template containing certain Lotus @functions whenever there are no data within the @function's specified range. It will also occur in cells that depend on other cells whose @functions evaluate to ERR. These displays are normal. A technique to eliminate ERR messages uses the Lotus @IF and @ISERR @functions. If you do not want ERR to appear in a cell, use the following formula:

```
@IF(@ISERR(@AVG(B9..E9)," ",@AVG(B9..E9))
```

2. Type the following formulae into the specified cells:

Cell	Formula
B22	@AVG(B9..E9)
C22	@AVG(B10..E10)
D22	@AVG(B11..E11)
E22	@AVG(B12..E12)
F22	@AVG(B13..E13)
G22	@AVG(B14..E14)

NOTE: You must type each of the formulae individually into the specified cells. You cannot issue the /*C*opy command to copy the formula from cell B22 to cells C22..G22 because the wrong cell designations would result. To eliminate ERR messages in the cells, use the @IF formulae described in Step 1.

3. Using the /***R***ange ***N***ame ***C***reate command, create the following ranges:

Range	Range Of Cells
STDX	A9..A14
STDF	F9..F14
STATS	C25
SLOPE	E32
INTERCEPT	F26
BC_CT	B39..B999
DNA_TRIO	H39..H50
FL_AVG	A21..G22
FL_CONC	A17..G18
FL_REC	A3..G4
FL_REP	B38..J50
FL_TRIO	B4..D4
RAW_STD	B9..E14
STAT_TB	C25..F33
TARGET	H39..J499
TECH_COMS	B5
TEMPLATE	A1..I51

The above ranges will be used as forms, field name header designators, tables, and @function ranges by future macro programs.

4. Issue a /***R***ange ***N***ame ***L***abel ***D***own command sequence and specify A3..G3 as the range. The A3..G3 range contains the field names Recno, ASSAY_TYPE, RUN_DT, BATCH, TECH, LOT_RGNT, and LOT_STDS. These names correspond to field names in BATCH.DBF and RESULTS.DBF. By issuing the command sequence, names corresponding to the field names in Row 3 will be assigned to corresponding cells in Row 4. Creating these named ranges will set up the template for programming that will be performed in future chapters. Using range-named cells in programs makes programs easier to read and follow. Perhaps more importantly, if you use range-named cells in your programs instead of cell addresses, you will not have to edit your program if there is a change in your template.
5. Into cell G9, type +(F9-$INTERCEPT)/$SLOPE (the formula for calculating the DNA concentration from the standard's average fluorescence value in cell F9 and the slope/intercept from the regression analysis).

   NOTE: Using a dollar sign before a cell name will allow you to "fix" the column and row coordinates of the cell. That is, if you were to copy a formula in a down direction without the preceeding dollar sign, the row coordinates for SLOPE and INTERCEPT would change by the number of rows that offsets the final cell from the original cell. However because a dollar sign is in front of $SLOPE and $INTERCEPT, they will be copied with no change in their cell designations.
6. Issue the /***C***opy command to copy cell G9 to cells G10..G14. (i.e., When prompted for the range to copy from, type G9 and press [RETURN]. When prompted for the range to copy to, type G10..G14 and press [RETURN].) The formulae will be copied to cells G10..H14.
7. Into cell F39, type +(C39*@AVG(D39,E39)-$INTERCEPT)/$SLOPE (the formula for calculating the DNA concentration from the dilution factor, the average of the fluorescence values in cells D39..E39, and the slope/intercept from the regression analysis).
8. Issue the /***C***opy command to copy from cell F39 to cells F40..F50.
9. Set the source for the print function to A1..G50. In Lotus 1-2-3, issue the /***P***rint ***P***rinter ***R***ange command sequence. In Symphony, issue the [SERVICES] ***P***rint ***S***ettings ***S***ource ***R***ange command sequence. When prompted for the range, type A1..G50 and press [RETURN]. Press [ESC]ape until you return to the spreadsheet.
10. Format cell C4 to long international date. For Lotus 1-2-3, use the /***R***ange ***F***ormat ***D***ate ***4***(Long Int'l) command sequence. For Symphony, use the /***F***ormat ***D***ate ***4***(Full Int'l) command sequence.
11. Store the file. To store the file in Symphony, issue the [SERVICES] ***F***ile ***S***ave command sequence. In Lotus 1-2-3, the command sequence is /***F***ile ***S***ave. When the program prompts for a name, type "DNA" and press [RETURN].

CONVERTING EXISTING TEMPLATES

If you have existing data acquisition and analysis templates and find that you must modify them to include database management properties, you can avoid re-typing the template by using the /*M*ove command to move cells. If necessary, you can insert or delete rows or columns. However, you must be careful when moving cells or ranges of cells. Moving cells often creates problems, especially with ranges of cells.

Before you begin the modification process, print out a table of the named ranges. Then, after changes have been made, you can compare the pre-modification table against the post-modification table to verify that all of the ranges still exist and that they contain appropriate cell assignments.

To make a range table, issue the /*R*ange *N*ame *T*able command. When prompted for the location of the table, press [ESC]ape and move the cell-pointer to a cell outside the area being used for the template. Press [RETURN] and the table will be created.

If changes are substantial, you may want to consider deleting all of the range names and re-creating the named ranges. Sometimes moving ranges causes overlaps in two or more ranges. That is, the same range of cells get more than one name. Overlapping ranges not only creates a potentially dangerous situation (changing the size or position of overlapping named ranges gives confusing and unpredictable results), but also ***substantially*** slows the performance of the template and macro programs using the template. To delete all ranges, issue the /*R*ange *N*ame *R*eset command.

Finally, if you have calculations in your template or macro programs that interact with the template, you should review the ranges in them. If you have used cell locations in your macro programs instead of named ranges, the cell locations will explicitly need to be updated.

TEST THE TEMPLATE

Now, take the time to see how the template works. Type the numbers shown in Figure 3-2 into cells B9 through E14 and A39 through E50. Next, perform a regression analysis on the standard curve.

If you are using Symphony, you will need to perform a preliminary task. The Lotus 1-2-3 regression function is permanently integrated into the Lotus 1-2-3 work environment and is available immediately. However, in Symphony the regression program is an add-in application program and must be attached before it can be used.

To attach the STAT.app program, press [SERVICES] *A*pplication *A*ttach, highlight STAT.app, and press [RETURN]. Press [ESC]ape until you return to the spreadsheet. The STAT.app program needs to be attached only once during a Symphony session.

To bring up the Lotus regression menu, issue the /*D*ata *R*egression command sequence (Lotus 1-2-3) or the /*R*ange *R*egress command sequence (Symphony). To compute the regression statistics, you must first define the X-range, Y-range, and Output-range or Lotus will issue an error message and abort the regression program. Choose the X-range, Y-range, and Output-range options in turn to specify the following:

Menu Choice	Range
X-range	STDX
Y-range	STDF
Output-range	STATS

After the appropriate specifications have been set, execute the regression analysis by choosing the *G*o menu option. Figure 3-2 shows the outcome of this action. The formulae and @functions in the template will automatically update after each number is added.

Actually, by typing in representative data and performing the regression analysis you have performed an even more important task—you have tested the template. ***Testing a template before you start programming is very important. If the template does not perform correctly alone, you cannot expect it to perform correctly when it is interacting with a program or with other templates.***

As a general rule, it is best to test templates as much as possible ***before*** going on to programming tasks. This action makes it easier to identify and track down errors. That is, if you wait until after you write your program, there will be many program lines and interdependencies in your template. These relationships will make it much more difficult to isolate errors. However, if you test as you build, you can verify that each step works correctly before you add another layer of complexity.

When you are finished testing the template, store it in a file called EXAMPLE. (Although you would not normally store a template with its data, the data in this template will be used as examples in future chapters of this book.)

To store the file in Symphony, issue the [SERVICES] ***F***ile ***S***ave command sequence. In Lotus 1-2-3, the command sequence is /***F***ile ***S***ave. When the program prompts for a name, type "EXAMPLE" and press [RETURN].

Then, erase the artificial data. In Symphony, issue the /***E***rase command. In Lotus 1-2-3, issue the /***R***ange ***E***rase command sequence. When prompted for the range to erase, press [ESC]ape, arrow over to cell B9, type a period (.), arrow over to cell E14, and press [RETURN]. Repeat the process to erase cells A39 through E50. These commands will return your spreadsheet to an empty template. Store the empty template again in the file named "DNA".

CREATE AND ERASE CELLS FOR MAXIMAL SPEED

There is another benefit to testing templates. The slowest process during data input is the process of creating new cells in a template. Once created, the process of filling the cells with data is relatively quick. Likewise, overwriting old data in cells with new data is a relatively slow process. The fastest way for programs to place data in cells is to have the cells already created, but empty. You can set up this condition by forcing Lotus to create the cells at the time that you make up your original template.

Thus, you fill the cells with "dummy" data during your testing session and then erase the data. Erasing existing cells clears previous data, but leaves the cell's memory space and attributes intact. This procedure is faster than creating and filling new cells and it is also faster than replacing existing data in cells.

If you have large numbers of identical cells in a matrix, you can use the /*D*ata *F*ill (Lotus 1-2-3) or the /*R*ange *F*ill (Symphony) command sequence to fill your template with numbers. Then you should erase the numbers and use the empty cells to receive your data.

Do not charge right out and create all of the cells in the spreadsheet. Even though the cells do not contain data, they still use up memory, meaning that you are likely to run out of memory and you may actually be slowing the performance of your spreadsheet if you create too many cells. It also means that you will be storing bigger files on your disk. Therefore, you should create only the cells that you plan to use in your template.

BACK UP YOUR WORK

I recommend storing a spreadsheet template (without data) as soon as you have all of its components in place and again after you have tested the template. In fact, it is a good idea to store it often during creation. That way, if you have a power failure or a computer lockup, you will not lose the work you have already completed. If you have a large macro program and spreadsheet, you may want to store it under different names and on several floppy disks.

Before you store a template, you should take a moment to restore any pertinent settings to their original default values. For example, take a minute to reset label alignment settings in the current spreadsheet. Recall that we set the label alignment to "center" for column headers. For macro programs, you will want to left-justify text. As before, the worksheet will automatically perform the justification. In Symphony, issue the /*S*ettings *L*abel-Prefix *L*eft command sequence. In Lotus 1-2-3, the equivalent command sequence is /*W*orksheet *G*lobal *L*abel-Prefix *L*eft.

USER INTERFACES AND REPORT FORMS

A user interface is the means by which a user can interact with a LIM system. The user inputs, edits, and views data via the user interface. Preventing mistakes at this level is very important because it is the one element that you cannot fully control through programming. Therefore, it is crucial that you design optimal user interfaces.

As a general rule, the easier it is for a user to interpret an interface, the less training that you will need to provide and the fewer mistakes that will be made. "User friendly" interfaces—interfaces that make a user's tasks as easy to discern and accomplish as possible—allow users with little specialized knowledge to use your LIM system. Interfaces can be made user friendly through orderly design.

User interfaces consist of two parts. The first part is, in actuality, just a template. The second part is a macro program that coordinates the updating of data in the template and transfer of data to other templates and/or database files. This section discusses the template design aspects of the user interface. Future chapters contain information on

```
--------O---------P---------Q---------R--------S--------T--------U----
 1  SID_FM 0,0
 2  DATE:         ________                     RECORD NUMBER:    _______
 3
 4                          SAMPLE IDENTIFICATION
 5
 6  SAMPLE BARCODE NUMBER:           _______
 7  SAMPLE NAME/DESCRIPTION:         _________________
 8
 9  DATE OF REQUEST:    _______      SAMPLE DELIVERY DATE:   ________
10  REQUESTER'S NAME:   _________    SAMPLE RACK BARCODE:    _______
11  DEPARTMENT NUMBER:  ___          STORAGE LOCATION:       ________
12  PROJECT CODE:       _____
13
14  ASSAYS REQUESTED:   ____________________________
15  COMMENTS:           __________________________
16
17                      SAMPLE STATUS:     ___
18
19      (PRESS [CONTROL][BREAK] WHEN DONE ENTERING DATA)
20
```

Figure 3-5

macro programming techniques for user interfaces.

There are basically two different types of user interfaces. You will probably need both types. They are

- templates that handle many records.
- templates that handle one record at a time.

Templates that handle many records simultaneously allow efficient viewing. More importantly, by containing many records a high degree of efficiency can be achieved for data input. For example, suppose a set of records contained repetitive data for one or more fields. By typing the data into the fields of the first record and copying the data to the corresponding fields of the other records, a great deal of typing could be avoided. The entire table could then be exported to a database file using the @BASE menu transfer utility.

Designing user interfaces to handle many records follows the same guidelines as the example template shown in Figure 3-1. In fact, you can often use templates designed for data handling as user interfaces and report forms and to meet multi-purpose template goals.

Single record templates are typically used when each record has unique information or when only a few records are being input. For example, when entering data into a database file containing information on lot numbers of reagents or sets of standards, a single record template is typically used.

Single record user interfaces follow many of the same rules as multiple record

```
---------Y--------Z--------AA-------AB-------AC--------AD-------AE-----AF-
 1  CERT_FM 0,0
 2  DATE:        ________                          RECORD NUMBER:     _______
 3
 4                           CERTIFICATE OF ANALYSIS
 5
 6  SAMPLE BARCODE NUMBER:           _______
 7  SAMPLE NAME/DESCRIPTION:         ________________
 8
 9  REQUESTER'S NAME:      _________               DEPARTMENT NUMBER:  ___
10  DATE OF REQUEST:       ________                PROJECT CODE:       _____
11
12
13  ASSAY TYPES REQUESTED:           ______________________________
14
15
16
17    ASSAY                    DATE      BATCH     ASSAY      PLATE
18     TYPE       RESULTS  PERFORMED  NUMBER   DILUTION   NUMBER    STATUS   TECH
19  ASSAY_TYPE  RESULTS    RUN_DT     BATCH   DIL_FCTR  PLATE_NUMR_STATUS  run_tech
20  _______    _______   _______   _______   _______   _______   _______   _______
21  _______    _______   _______   _______   _______   _______   _______   _______
22  _______    _______   _______   _______   _______   _______   _______   _______
23  _______    _______   _______   _______   _______   _______   _______   _______
```

Figure 3-6

interfaces. When designing single record interfaces, you must try to balance two opposing objectives:

- Try to keep the entire template on one screen so that the form does not have to be scrolled to view all of the data in the template.
- Try to keep "clutter" to a minimum. That is, try to keep the amount of information on the screen to a minimum.

One way to reach a balance is by grouping related pieces of information in one area of the template. For example, Figure 3-7 shows a template designed for the input of data relating to a batch. Fluorescence values for the standard curve are grouped together as curve fitting parameters. By doing so, parallel columns of data can be created and cells can be placed in close proximity of each other without the template appearing cluttered.

Considering the above objectives, creating your own user interface is essential. You could alternatively create a macro program that could easily access and use the @BASE user interface for data input, editing, viewing, etc. However, the @BASE user interface is limited because data are presented either in columns or rows. These formats tend to make the display lack organization in terms of related information. Furthermore, @BASE uses field name acronyms in the user interface. Acronyms are often confusing, especially

```
---------AV--------AW-------AX-------AY--------AZ--------BA-------BB---
 1  BCH_FM 0,0
 2  DATE:          ________                        RECORD NUMBER:   ________
 3
 4                              ASSAY BATCH DETAILS
 5
 6  ASSAY TYPE:             _________              BATCH NUMBER:    __
 7  DATE PERFORMED:         ________               TECHNICIAN:      ____
 8
 9  REAGENT LOT #:          ________
10  STANDARDS LOT #:        ________
11
12            *****AVERAGE FLUORESCENCE FOR STANDARDS*****
13  ________ ________           ________ ________
14  ________ ________           ________ ________
15  ________ ________           ________ ________
16
17            *****CURVE FITTING PARAMETERS*****
18  SLOPE:     ________          STANDARD ERROR Y ESTIMATE:         ________
19  INTERCEPT:________           R SQUARED:                         ________
20
21  COMMENTS:  _____________________               FINAL REVIEW DATE: ________
22
23                       *****SAMPLE DATA*****
24
25    Recno    BARCODE DIL_FCTR RAW_DATA1 RAW_DATA2  RESULTS PLATE_NUM
26   ________ ________ ________ ________ ________ ________ ________
27   ________ ________ ________ ________ ________ ________ ________
28   ________ ________ ________ ________ ________ ________ ________
29   ________ ________ ________ ________ ________ ________ ________
30   ________ ________ ________ ________ ________ ________ ________
31   ________ ________ ________ ________ ________ ________ ________
```

Figure 3-7

to new users.

Because of the limitations noted above, the @BASE interface may be prone to mistakes by new users. Therefore, future chapters will show you how to create programs to implement your own user interfaces. The remainder of this section shows you how to create templates for user interfaces.

To create templates for your LIM system, start by reviewing the lists created in Chapter 2. You will need enough templates so that each field name on the lists appears in at least one template. Essentially, you will need at least one template for each database file in your system. The example system requires templates to service the SAMPLES.DBF, RESULTS.DBF, BATCH.DBF, REAGENTS.DBF, and STANDARDS.DBF database files.

Figures 3-5 through 3-10 show user interfaces designed for the example LIM system presented in this book. As you can see, the interfaces service each database file. In some instances, the interfaces contain fields not appearing in the database files serviced by the

```
--------AK-------AL-------AM-------AN-------AO-------AP-------AQ---
 1  SRD_FM 0,0
 2  DATE:      ________                    RECORD NUMBER:    ________
 3
 4                    SAMPLE RESULTS DETAIL
 5
 6  SAMPLE BARCODE NUMBER:   ________
 7  SAMPLE NAME/DESCRIPTION: ____________________
 8  REQUESTER'S NAME:  ________            DEPARTMENT NUMBER:________
 9  DATE OF REQUEST:   ________            PROJECT CODE:     ________
10
11  ASSAY TYPE:        ________            DILUTION FACTOR:  ________
12  RESULTS:           ________
13
14                   *****BATCH DETAILS*****
15  RUN DATE:          ________            PLATE NUMBER:     ________
16  BATCH:             ________            TECHNICIAN:       ________
17  DUPLICATE #1:      ________            STD LOT NUMBER:   ________
18  DUPLICATE #2:      ________            RGT LOT NUMBER:   ________
19
20                 ****STANDARD CURVE DETAILS****
21          SLOPE:     _______  INTERCEPT:       _______
22   STD ERR Y EST:    _______  R SQUARED:       _______
23    STD ERR COEF:    _______
24
25          COMMENTS:__________________________
26
```

Figure 3-8

templates. These fields are fields that are accessible from other database files using the join feature of @BASE. Note that the templates in the figures contain only text directing attention to input cells. To create the forms:

1.) Type the text as it appears in the figures. Create all of the templates into the spreadsheet created for Figure 3-1. Use Figures 3-5 through 3-10 as guidelines for cell locations.

2.) There are several styles that can be used for template input lines. One method is to place a backslash and an underline (_) in the appropriate cells. By doing so, underlines will appear across the entire cell. When longer underlines are needed, the backslash-underline can be copied to adjacent cells. Another method is to repeatedly type underline characters. Perhaps a more meaningful method to the user is to coordinate the line length with the length of the field in the database. Figures 3-5 through 3-10 were constructed according to this method. These figures use the @REPEAT @function to place lines of specific length into the templates. For example, according to the definition in Figure 3-3, ASSAY_TYPE has a length of 10 characters. Therefore, the following formula

```
-------BF------BG----BH----BI-----BJ-BK--BL-------BM----BN----BO-------BP-------
25  0,0 RG_FM
26  DATE:       ________                    RECORD NUMBER:      _______
27
28                                  ASSAY REAGENTS
29
30              LOT NUMBER:         __________________
31
32  REAGENT NAME:   __________________
33
34  REAGENT DESCRIPTION:            ______________________
35
36
37  DATE OF PREPARATION:            ________    EXPIRATION DATE:_______
38  SHELF LIFE:                     _______     EXPIRATION FLAG:______
39
40
41                    *****LOT NUMBERS OF COMPONENTS*****
42              1:  __________________              3:  __________________
43              2:  __________________              4:  __________________
44
45                  (PRESS [CONTROL][BREAK] WHEN DONE ENTERING DATA
```

Figure 3-9

was placed into cell AC15 of Figure 3-6:

@REPEAT("_",10)

Using the @REPEAT method is a very useful tool to help you lay out templates because underlines representing the length of the fields being used in forms help expose clutter and overlapping cells. The method is also often helpful to the user because the numbers of acceptable characters for fields are readily apparent.

3.) Set column widths so that text prompts in the templates do not overlap with data cells. For Lotus 1-2-3, the command sequence is /*W*orksheet *C*olumn *S*et-Width. For Symphony, the command sequence is /*W*idth *S*et.

4.) The macro programs interacting with the user interface and report form templates will need a number of named ranges. These named ranges will be used as definitions for the forms, field name header designators, tables, and @function ranges. Issue the /*R*ange *N*ame *C*reate command to create the ranges shown in Figure 3-11.

5.) Date format the following cells to long international date. For Lotus 1-2-3, use the /*R*ange *F*ormat *D*ate *4*(Long Int'l) command sequence. For Symphony, use the /*F*ormat *D*ate *4*(Full Int'l) command sequence.

```
-------BF------BG----BH----BI-----BJ-BK--BL-------BM----BN----BO-------BP---BQ
  1  0,0 STD_FM
  2  DATE:________                    RECORD NUMBER:        ________
  3
  4                         STANDARDS INFORMATION
  5
  6       LOT NUMBER:   _________
  7
  8      NAME OF SET:   _________
  9  SET DESCRIPTION:   _________
 10
 11  PREPARATION DATE:  _________           EXPIRATION DATE:________
 12  SHELF LIFE:        _________           EXPIRATION FLAG:_______
 13  TECHNICIAN:        _________
 14
 15            LOT NUMBER        [CONC]           LOT NUMBER        [CONC]
 16      0  ____________________ _______    4  ____________________ _________
 17      1  ____________________ _______    5  ____________________ _________
 18      2  ____________________ _______    6  ____________________ _________
 19      3  ____________________ _______    7  ____________________ _________
 20          (PRESS [CONTROL][BREAK] WHEN DONE ENTERING DATA)
```

Figure 3-10

P2, Q9, U9, Z2, AA10, AA20..AA30, AW2
AX7, AL2, AM9, AM15, AY2, BA13, BA14
BD14, BG2, BI11, BI12, BO11

6.) Save the spreadsheet under the name "DNA". To store the file in Symphony, issue the [SERVICES] ***F***ile ***S***ave command sequence. In Lotus 1-2-3, the command sequence is /***F***ile ***S***ave. When the program prompts for a name, type "DNA" and press [RETURN].

USER INTERFACE GUIDELINES

When creating your own user interface templates, consider the following guidelines:

- Look at the lists of databases created as a result of Chapter 2 and create user interfaces for each database.
- Look at the lists of reports and create report forms to cover all items on the list.
- Optimally, you should strive to design each user interface template to be multi-purpose in that the template would be re-usable for all phases of data handling (inputing, editing, printing, and data display).
- Set column widths so that text prompts in the templates do not overlap with data cells. Entering data into data cells will prevent the overlapping portion of the

Range	Range Of Cells
BCH_FM	AV1..BB22
BCH_HDR	AV25..BB25
BCH_TAB	AV26..BB106
B_SD	AV25..BB25
CERT_FM	Y1..AF15
CERT_HDR	Y19..AF19
CERT_TAB	Y20..AF30
RG_FM	BF25..BP45
STD_FM	BF1..BP20
SID_FM	O1..U19
SRD_FM	AK1..AR25

Figure 3-11

prompt from being displayed and will cause confusion.

- Try to group similar data fields together. Use parallel columns if necessary.
- If possible, separate groups of data with open rows.
- Be keenly aware of the user. The user interface is what the user sees and interacts with. If the interface is confusing to the user, the user may make mistakes that could prove costly.
- Try to keep the length of entries in the fields as short as possible. Short field entries conserve memory and reduce keying time and errors.
- Try to keep user interface templates compact and on one screen so the user does not have to scroll to review information.
- If the system has more than one user interface, position the interfaces side-by-side, across columns. Do not stack them on top of each other. That way, if you need to add information, you can expand the template by adding rows below the template. Perhaps more importantly, it is easier to customize column widths in templates when they are not stacked.
- Try to keep templates less than 82 columns wide. This width will allow printing of the template without having to place the printer in compressed mode.

When input is needed, you can implement several systems that help the user to step through the input process. The following is an overview of some systems that you can use for this purpose:

- Create a program with {GOTO} and {?} commands that move the cell pointer sequentially through the cells to be updated and wait until a carriage return has been pressed before proceeding to the next cell.
- Create a program with {GETLABEL} and {GETNUMBER} commands to

prompt the user on the second line of the control panel, pause the program until input is made, and then place the data into a specified cell.

- Create menus to prompt for input, such as assay type.
- Include an @NOW @function to automatically set the current date.
- Perform automatic sample numbering by creating a macro program to search a database for the highest Recno, add one, and place the new record number into the entry form as the sample number.

These programming techniques allow you to limit user input to only those fields on the template that are appropriate to the function being performed. For example, using the above techniques will allow you to restrict access to cells containing sample results data.

SUMMARY OF WHAT HAS BEEN ACCOMPLISHED

In this chapter, you learned how to design and create an organized skeleton to hold your data. This superstructure will facilitate manipulation and presentation of data. The multi-purpose assay template created in this chapter made extensive use of matrices to efficiently integrate data acquisition, data analysis, user input, transfer of data to database files, and report generation. Multi-purpose templates conserve memory and increase program speed and reliability.

You also learned of the importance of placing data in tables that conform to @BASE requirements for menu-based transfer of data between spreadsheets and database files. The tables need not contain all of the fields specified in a database, but must have a two-part structure consisting of a header of field names, followed by a body of data. There can be no blank rows or columns within the table. For some applications, it is important to include a preceding Recno field in the table.

In anticipation of macro programs, you learned how to create named cells and ranges to hold data from experiments. These named cells and ranges allow you to use names in the commands of data reduction programs; the programs are therefore easier to read. These named cells and ranges also serve to "marry" the template to a macro program.

One of the most important concepts presented in this chapter is the importance of reviewing the overall design of your LIM system. After you have designed your templates, it is very important to thoroughly review all of the field name lists and flow diagrams created in Chapter 2. Review them to ensure that all data needed to update the relevant databases are contained in the template. You should also review the lists for correctness and completeness. The creation of templates often exposes errors in the lists. In fact, generating templates is a valuable tool for ensuring the appropriateness of field name lists.

WHAT'S NEXT?

The next chapter will show you how to install and configure @BASE and the @BASE Option Pac and how to create database definitions.

4

Creating and Modifying Database Definitions

After you have completed the steps presented in Chapter 2 to determine the fields of information that your databases will need to contain and how the information in various database files will relate, you can begin to create the databases. The add-in software programs, @BASE™ and @BASE Option Pac, make building and joining databases easy. When you build new databases, you must create ***database definitions*** before you can enter data. For each field in a new database, a database definition contains specifications for the field name, field type, field length, and number of decimal digits. Thus, a database definition tells the @BASE database management system how the fields are organized for each record. There are three ways to create a database definition:

- Manually type the definition using an @BASE ***F***ile ***D***efine menu choice.
- Translate the definition from a Lotus spreadsheet template using an @BASE ***U***tility ***F***ile ***T***ranslate menu choice.
- Extract the definition from an existing database file using an @BASE ***F***ile ***E***xtract menu choice.

After you have defined a new database, you can begin entering data. There are four ways to add data to a database:

- Manually type data into the database using an @BASE ***D***ata ***E***dit, ***D***ata ***F***orm, or ***D***ata ***B***rowse menu choice.
- Export data from a spreadsheet template to a database file using an @BASE ***D***ata ***T***ransfer ***E***xport or ***O***ptions ***I***ndex ***T***ransfer ***E***xport menu choice.
- Extract data from an existing database file created by Ashton-Tate® dBASE III®,

dBASE III Plus®, or @BASE.

- Transfer data to the database file from a spreadsheet template using @BASE @DB and @NDX @functions placed in the commands of macro programs.

This chapter shows you how to install and configure @BASE and the @BASE Option Pac and how to create database definitions. Chapter 5 shows you how to join database files to create relational databases.

You can still make changes to the database definitions once you have defined a database and entered data. Database definitions can be modified at any time. You can add fields, delete fields, change field names, change field types, lengthen or shorten fields, etc. @BASE automatically converts existing data to comply with the modified definitions.

THE @BASE PROGRAMS

@BASE and the @BASE Option Pac consist of a series of interrelated programs that work together to provide you with powerful database management capabilities. @BASE lets you build and manage sophisticated database applications from within the familiar Lotus environment. Lotus provides the tools for data acquisition and analysis and @BASE provides the access speed and nearly unlimited data archiving capacity that you will need for your LIM system. The @BASE programs that you will be installing in the next sections are:

- @BASE.ADN (Lotus 1-2-3) and @BASE.APP (Symphony) are the main @BASE programs.
- @BASEFUN.ADN (Lotus 1-2-3), @BASEFUN.APP (Symphony, 1.2 and 2.0) and BASEOPTF.APP (Symphony 2.2) are companion programs for @BASE that provide special database @functions.
- @BASE.TXT is a file that contains text messages for @BASE.
- @BASEUTL.ADN (Lotus 1-2-3) and @BASEUTL.APP (Symphony) are coadd-ins that provide several utilities for @BASE.
- @BASEOPT.ADN (Lotus 1-2-3) and @BASEOPT.APP (Symphony) are Option Pac programs that contain @BASE's indexing and relational capabilities.

Before you can use the @BASE programs, you must first install them. The next section reviews some important memory information that you should consider before installing and using @BASE and the @BASE Option Pac. The section following provides you with information on how to install both @BASE and the @BASE Option Pac.

CONFIGURING YOUR SYSTEM FOR EFFICIENT MEMORY USAGE

@BASE database files are stored on disk. When working with large quantities of data in a database, storing the database on disk rather than in memory allows you to store many more data records. Lotus, @BASE, and the @BASE Option Pac are all stored in your computer's conventional memory, occupying substantial amounts of this memory, thereby limiting the size of the spreadsheet you can create before you obtain a "Memory Full" message. Therefore, it is important for you to modify your system to try to obtain as much memory as possible. Chapter 1 contained some specific recommendations on how to configure your system to make as much conventional memory available as possible. Three of the items on the list need to be implemented as part of this chapter. The three items directly relating to this chapter are:

- The DOC, FORM, and COMM environments of Symphony are not normally needed and can be unloaded with a program called "EXTRA K". This program is available from Lerman Associates (12 Endmoor Road, Westford, MA, 01886, 508-692-7600) or directly from the author using the order form at the end of this book. After Symphony, @BASE and the @BASE Option Pac have been loaded, there can be as few as 9,600 cells remaining for data (depending on Release and data type). Removing the three environments with EXTRA K will provide up to about 3,000 more cells for spreadsheet use in systems without expanded memory and up to about 21,000 more cells in systems with expanded memory. If more memory is needed, the Lotus HELP system can also be removed. Removing the HELP system will provide up to about 300 additional cells.

  Because EXTRA K is a Symphony add-in that reorganizes memory and monitors the Symphony resources used by other add-in application programs, ***EXTRA K must be the first application program attached and invoked. Furthermore, there must be no data in any spreadsheet prior to invoking EXTRA K.***

  To install EXTRA K, simply copy the contents of the distribution disk to the same directory as the Symphony program files. Start Symphony with no attached add-ins and no spreadsheet data. Determine which EXTRA K file to use. The EXTRA K add-in you need to attach depends on whether you are using expanded memory. If you are unsure whether your system uses expanded memory, issue a [SERVICES] ***S***tatus command sequence. The status display will include an entry for expanded memory. If "(None)" is displayed, then there is no expanded memory and you should use EXTRAK.APP. Otherwise Symphony will display the total available expanded memory and you should use XEXTRAK.APP.

  To ensure the proper environment at startup, configure Symphony to auto-load the appropriate EXTRA K add-in program. Auto-loading the add-in module will decrease mistakes, confusion, and program errors. The first steps to configuring Lotus for auto-loading is to access the Setup menu. Issue the [SERVICES], ***C***onfiguration, ***O***ther, ***A***pplication command sequence. The display in Figure 4-1 will appear. Because EXTRA K must be the first add-in, select the ***S***et and ***1*** options. Select the appropriate EXTRA K program from the menu and press [RETURN]. Then Select ***N***o for auto-invoke.

  EXTRA K will then invoke and you will be presented with a screen of options. From the menu, choose ***D***OC, ***F***ORM, ***C***OMM and ***H***elp. Select

```
Select an add-in to automatically attach, and attach it
Set    Cancel    Quit
________________________________________________________

1:                      Auto-Invoke: No
2:                      Auto-Invoke: No
3:                      Auto-Invoke: No
4:                      Auto-Invoke: No
5:                      Auto-Invoke: No
6:                      Auto-Invoke: No
7:                      Auto-Invoke: No
8:                      Auto-Invoke: No
________________________________________________________
```

Figure 4-1

Update and then ***S***tartup-directory. EXTRA K will write a configuration file to disk that will automatically set the environments to be disabled by EXTRA K each time EXTRA K is invoked.

Next, choose ***P***roceed ***Q***uit and ***U***pdate to update the Lotus configuration file. Press [ESC]ape until you have returned to the spreadsheet. Thereafter, the EXTRA K add-in module will be automatically loaded each time you start Symphony.

Exit Symphony. Using your word processor, either create a file named AUTOEXEC.BAT in your root directory with the following line or add the line to your existing AUTOEXEC.BAT file:

X:\path\EKAUTO X:\path -ON

where, X: is the drive and "path" is the directory containing your Symphony files.

Re-boot your computer. This set of commands will execute the program named "EKAUTO" whenever your computer is booted. EKAUTO is a program that tells EXTRA K to bypass its initialization menu when it attaches. Bypassing the menu is useful because it does not require the user to supply input at the beginning of a session.

- The Undo feature of Lotus 1-2-3, Release 2.2, requires a lot of memory (about 163,400 bytes). In fact, this amount of memory is so large that it is possible to receive a "Memory full" message immediately after @BASE and @BASE Option Pac are loaded. It is therefore important to disable the Undo feature of Lotus Releases incorporating the feature. Do so by issuing a /***W***orksheet ***G***lobal ***D***efault ***O***ther ***U***ndo ***D***isable ***U***pdate ***Q***uit command sequence. Disabling Undo will free about 6,000 cells worth of memory in systems with only conventional memory and about 42,000 cells in systems with expanded memory. Upon returning to the spreadsheet, you may note that the "Memory full" message will still be displayed. This message indicates that all of the memory available has not been released back to the pool. To release the memory, exit and re-enter Lotus 1-2-3.

- Remove any memory-resident (Terminate-Stay Resident, TSR) programs not in use. Programs such as SmartNotes and Sidekick are loaded into memory before Lotus is loaded, leaving less room for spreadsheets. Unloading these programs before entering Lotus will make more memory available. To do so, remove the lines in the CONFIG.SYS file specifying the programs and re-boot your computer.

CONFIGURING YOUR SYSTEM FOR OPEN FILES

If, when using this system, you receive a "Cannot open file," "Locked file," or "Cannot create file" error message, check your CONFIG.SYS file for a FILES command to ensure that the command is present and specifies at least 20 files. If FILES=20, then a maximum of about 14 files are available because Lotus and @BASE normally maintain up to about 6 open files. Because your system may utilize up to 14 open files, use your word processor to change the FILES statement to at least 20 and re-boot your computer. (The maximum number of files is 255. However, DOS allows no more than 20 open files to each software application. So, increasing the number to more than 20 files will not increase the number of files available to Lotus unless you have TSR files running that also have open files. If your system exceeds the 14 open file limit, then you will need to review the system and try to combine database files or close open index or database files to decrease the number of open files to an acceptable limit.)

INSTALLING @BASE

With Lotus 1-2-3, Releases 2.0 and 2.01, the Lotus Add-in Manager needs to added to the ".SET" driver file for your spreadsheet. For Lotus 1-2-3, Releases 2.0, 2.01, and 2.2, @BASE and @BASE Option Pac modules need to be copied to the subdirectory containing Lotus files.

Once the above have been accomplished, @BASE can be made to automatically load each time Lotus 1-2-3 is started. Then, when you want to use the @BASE menu, you press a key combination (such as [ALT]-[F9]) to activate the module's menu. The {APP} command is a Lotus 1-2-3 macro command that is equivalent to pressing the [ALT] key combinations that invoke add-ins. You can use the {APP} command to execute one of the module's menu choices from within your macro. {APP3} for example is equivalent to pressing [ALT]-[F9].

With the @BASE Release for Symphony, the modules act exactly like other Symphony add-in application programs. That is, akin to the DOS.app and STAT.app programs shipped with Symphony, you must attach @BASE and the @BASE Option Pac application programs to Symphony before you can use them. With Symphony Releases 1.2 and 2.0, this procedure adds the name of the modules to the SHEET menu. (The SHEET menu key is [F10], or the forward slash, "/".) Thereafter, just choose the module's name from the SHEET menu and the application program will be invoked. Likewise, you can issue a '/, {MENU}, or {M} macro command in your program and invoke your add-in to have it perform specific tasks.

Symphony Release 2.2 works slightly differently. After the modules are attached, you must change environments before you can invoke @BASE. @BASE does not appear on the SHEET menu of the spreadsheet environment. To change environments, press [SHIFT]-[F10] and choose ***B***ase. To access the environment type menu from a macro program, issue a {TYPE}B command.

Installing @BASE for Lotus 1-2-3

The Lotus Add-in Manager needs to be added to your Lotus 1-2-3 Release 2.0 or 2.01 driver set. (The Add-in Manager is a permanent part of Lotus 1-2-3, Release 2.2.) A Lotus driver set is a file that brings together all of the software programs that drive the pieces of hardware used by Lotus. For example, a driver set includes software drivers for your monitor and printer. After you have completed this section of the book, your 1-2-3 driver set will also hold the Add-in Manager that allows you access to @BASE.

Personics Corporation has made the addition of the Add-in Manager and copying of the @BASE files to the Lotus 1-2-3 subdirectory almost trivial by including an installation program on their disks. The name of the program is "Install". Using the installation program depends on your version of @BASE.

For @BASE, the current drive must be the a: drive. Therefore, place the @BASE disk into drive a:, type a:, and press [RETURN]. Then type INSTALL and press [RETURN].

For the @BASE Option Pac, @BASE must first be installed. For the Option Pac, the current drive and directory must contain your 1-2-3 program files. For example, if your 1-2-3 program files are in drive c: and subdirectory \123, type c:, press [RETURN], type cd\123, and press [RETURN]. Place the @BASE Option Pac disk into drive a:, type A:INSTALL, and press [RETURN].

The installation programs will ask whether you want to auto-load the add-ins. Answer "no" to all of the questions. Once you have installed @BASE, you will not have to repeat the procedure again.

Installing @BASE for Lotus Symphony

The procedure is slightly different for Symphony. Because the @BASE database modules for Symphony are add-in application programs, they need to be copied into the subdirectory containing your Symphony program.

With Symphony Release 2.2, @BASE is installed as part of the installation process of Symphony. However, the @BASE Option Pac needs to be installed. For previous

Lotus 1-2-3 Version

```
A1:

Set    Cancel    Update    Quit
Select an add-in to automatically attach, and attach it
________________________________________________________

1:                     Auto-Invoke: No        KEY:
2:                     Auto-Invoke: No        KEY:
3:                     Auto-Invoke: No        KEY:
4:                     Auto-Invoke: No        KEY:
5:                     Auto-Invoke: No        KEY:
6:                     Auto-Invoke: No        KEY:
7:                     Auto-Invoke: No        KEY:
8:                     Auto-Invoke: No        KEY:
________________________________________________________
```

Symphony Version

```
Select an add-in to automatically attach, and attach it
Set    Cancel    Quit
________________________________________________________

1:                     Auto-Invoke: No
2:                     Auto-Invoke: No
3:                     Auto-Invoke: No
4:                     Auto-Invoke: No
5:                     Auto-Invoke: No
6:                     Auto-Invoke: No
7:                     Auto-Invoke: No
8:                     Auto-Invoke: No
________________________________________________________
```

Figure 4-2

Releases of Symphony, both @BASE and the @BASE Option Pac need to be installed.

Personics Corporation has made these activities almost trivial by including installation programs on their disks. The name of the program to use depends on your add-in and Release of Symphony. For example,

- Symphony 1.2, 2.0; @Base: BSETUP
- Symphony 1.2, 2.0; @BASE Option Pac: INSTALL

- Symphony 2.2; @BASE Option Pac: OPSETUP

For all installation programs, the current drive must the a: drive. Therefore, place the @BASE disk into drive a:, type a:, and press [RETURN]. Then type the name of the appropriate installation program and press [RETURN]. @BASE must be installed before the @BASE Option Pac can be installed.

The installation program will ask whether you want to auto-load the add-ins. Answer "no" to all of the questions. Once you have installed @BASE, you will not have to repeat the procedure again.

CONFIGURING LOTUS TO AUTO-LOAD @BASE

To ensure the proper environment at startup, configure Lotus to auto-load the @BASE add-in programs. Auto-loading the add-in modules will decrease mistakes by avoiding confusion and program errors. The first steps to configuring Lotus for auto-loading depends on your Release. After you have accessed the Setup menu, selecting add-in programs for auto-loading is independent of Release.

Begin by accessing the Lotus Setup menu. The following are the keystrokes to use:

- Lotus 1-2-3, Release 2.0, 2.01: [SHIFT]-[F10], *S*etup.
- Lotus 1-2-3, Release 2.2: /, *W*orksheet, *G*lobal, *D*efault, *O*ther, *A*dd-In.
- Symphony 1.2, 2.0, 2.2: [SERVICES], *C*onfiguration, *O*ther, *A*pplication.

The upper display shown in Figure 4-2 will appear for Lotus 1-2-3. The lower display will appear for Symphony. (NOTE: EXTRA K would appear on the Symphony display if it had been previously installed.)

After you have accessed the Setup menu, the keystroke sequence is independent of the Release of Lotus software you are using. Repeatedly choose the *S*et option from the menu to specify the following:

- *S*et, *1*, @BASE, *9*, *N*o
- *S*et, *2*, @BASEFUN (Lotus 1-2-3 or Symphony 1.2 or 2.0) or BASEOPTF (Symphony 2.2)
- *S*et, *3*, @BASEOPT, *N*o, *N*o

Note that:

- Lotus 1-2-3 allows you to specify 9 to allow you to use [ALT]-[F9] or [SHIFT]-[F9] to invoke @BASE. The @BASE Option Pac is accessed directly from the @BASE menu.
- Symphony, Release 2.2, add-ins do not have "@" signs as part of their names.
- The @BASEUTL.ADN and @BASEUTL.APP programs are ***not*** configured for automatic loading. These programs are used only during system setup and

consume memory. Therefore, they are attached only when needed.
- BASEFUNC is ***not*** used for Symphony 2.2.

Next, choose *U*pdate to update the configuration file and press [ESC]ape until you return to the spreadsheet. Thereafter, the three add-in modules will be automatically loaded each time you start up your Lotus program.

If you inadvertently configure your system to auto-load BASEFUNC for Symphony, 2.2, you must remove BASEFUNC. BASEFUNC contains a subset of @functions that are redundant with the @functions provided in BASEOPTF. This redundancy can cause Lotus to become confused, slows execution speed, and wastes memory. You cannot cancel the auto-load from the Lotus [SERVICES] *C*onfiguration menu because of a bug in Symphony. To do so, you must use the REM_ATB program that is supplied on the @BASE Option Pac disk for Symphony 2.2.

Likewise, if it becomes necessary to remove @BASEFUN from Symphony 1.2 or 2.0, the equivalent program is called DEL_ATB and is on the @BASE disk.

CONFIGURING @BASE

You can customize @BASE's global settings to tailor a system to your laboratory's preferences and save the default settings in a file. The default settings will be automatically retrieved by @BASE each time you begin a Lotus session.

Begin the process by accessing @BASE. The keystrokes to access @BASE depend on the Release of software you are using:

- Lotus 1-2-3, 2.0, 2.01, 2.2: [SHIFT]-[F9].
- Symphony 1.2, 2.0: [F10], ***@***Base
- Symphony 2.2: [SHIFT]-[F10], ***B***ase, [F10]

After you have issued the appropriate keystrokes, you will be presented with @BASE's main menu. From the main menu, select *S*ettings. You will be presented with a menu that will allow you to change the settings. The following is a description of each menu choice:

- The *D*eleted setting determines whether @BASE uses or ignores records marked for deletion. For example, when you review, edit, or transfer data, @BASE will act as if records marked for deletion did not exist if *D*eleted is set to *I*gnore. To avoid confusion, the default, or preferred setting, is to skip deleted records.
- The *C*ase setting determines whether @BASE is case-sensitive with character strings. If you set @BASE to be case-sensitive, all of the commands that deal with strings or string functions will treat upper-case characters differently from lower-case characters. If *C*ase is set to *I*gnore-case, @BASE will treat upper-case and lower-case the same. For example, if *I*gnore-case is set, assay_type and ASSAY_TYPE are treated as identical. The default, or preferred

setting, is to ignore case.

- The ***F***ile-lock setting can be used to make @BASE "read only". This feature is useful for providing users with easy access to information in database files without giving them the ability to alter data. Thus, you can prevent novice users from changing database files and unwittingly losing data. If this setting is set to ***L***ock, @BASE can read any database file, but cannot add or change any data or database definition. You can also use this feature to lock out editing abilities on remote nodes of Local Area Networks so that only the originating laboratory can edit data. The default setting allows writing to database files. (Not available in Symphony, Release 2.2. With Symphony 2.2, you would lock out editing by opening files as "***R***ead-only," rather than "***E***xclusive".)
- The ***Y***ear setting is used to specify 2-digit or 4-digit years in dates. @BASE stores dates on disk in the dBASE III YYYYMMDD format. Therefore, date fields are always eight characters long regardless of the current ***Y***ear setting. This setting affects the way that dates are input as well as the way in which they are displayed. Dates used in criteria expressions must conform to this setting. If ***Y***ear is set to 2-digit, @BASE displays only the last 2 digits of the year when displaying dates. New dates being entered are assumed to be in the 19xx format. If ***Y***ear is set to 4-digit, @BASE accepts years in the range from 1900 to 2099. All four digits of the year are displayed whenever a date field is displayed. All four digits must be input when entering a new date. The default setting is 2-digit years. (Not available in early Releases of @BASE for Lotus 1-2-3.)
- The ***B***lanks setting tells @BASE how to treat blank fields in a database. This setting affects date, numeric, and logical fields. Blanks can either be treated as zeroes or as NA (not available). When set to ***N***A, a blank in a date or numeric field indicates that a value is not available and any mathematical computations using that field will also have a result of NA. The default setting is to treat blanks as zeroes. However, the preferred method is ***N***A because zeroes will cause unnoticed errors in calculations. If a calculation returns NA, then the user will at least know that something has gone wrong. (Not available in early Releases of @BASE for Lotus 1-2-3.)

After you have entered your custom specifications, be sure to use the ***S***ave or ***U***pdate option on the menu so save the settings. The settings are saved to a configuration file called @BASE.CFG or BASE_SYM.CFG. The saved settings are automatically retrieved by @BASE when it is loaded into memory.

CREATING A DATABASE DEFINITION USING *F*ILE *D*EFINE

The first step in creating a new database file is to define the file's fields. To begin this process, access the @BASE menu using the keystrokes specified in the previous section. After you have issued the appropriate keystrokes, you will be presented with @BASE's main menu. Choose ***F***ile, then ***D***efine. ***F***ile ***D***efine allows you to define the fields.

Assay Batch Database (BATCH.DBF)

Identification Fields

Record number (from @BASE)	Recno:N:8:0
Assay type	ASSAY_TYPE:C:10
Date assay performed	RUN_DT:N:5:0
Batch number	BATCH:N:3:0
Lot number of standards used	LOT_STD:C:20
Lot number of reagents used	LOT_RGNT:C:20

Information Fields

Technician performing assay	BCH_TECH:C:5
Average response for standard #0	AVG_S0:N:8:4
Average response for standard #1	AVG_S1:N:8:4
Average response for standard #2	AVG_S2:N:8:4
Average response for standard #3	AVG_S3:N:8:4
Average response for standard #4	AVG_S4:N:8:4
Average response for standard #5	AVG_S5:N:8:4
Average response for standard #6	AVG_S6:N:8:4
Average response for standard #7	AVG_S7:N:8:4
Curve fit-A Parameter/Slope	CURV_A:N:8:4
Curve fit-B Parameter/Intercept	CURV_B:N:8:4
Curve fit-C Parameter	CURV_C:N:8:4
Curve fit-D Parameter	CURV_D:N:8:4
Standard Error of Y Estimate	STD_Y:N:8:4
Correlation Coefficient Squared	R_SQR:N:6:4
Number of Observations	N_OBS:N:3:0
Degrees of Freedom	DF:N:3:0
Standard Error of Coefficient	STD_C:N:7:4
Average response for control #1	AVG_C1:N:8:4
Average response for control #2	AVG_C2:N:8:4
Date of final review by supervisor	FREV_DTE:D:9
Technician comments	TECH_COMS:C:21

Figure 4-3

Each field has up to four elements that must be specified. They are:

- The name of the field.
- The field type (character, numeric, logical, or date).
- The field length (number of characters or digits allowed).
- The number of decimal places (if field is numeric).

The database that will be created in this section is the BATCH database from Chapter 2 (Figure 2-8). It is shown again in Figure 4-3 for your convenience. The new database file will be called BATCH.DBF. When prompted for the name, type BATCH and press [RETURN]. In response, @BASE will display a screen that prompts you through the field definition process. When creating a database definition, you are prompted for each of the above-mentioned elements that describe the fields.

Reagents Database (REAGENTS.DBF)

Identification Fields

Record Number (From @BASE)	Recno:N:8:0
Lot number of entire set	LOT_RGNT:C:20

Information Fields

Reagent name	RGNT_NAME:C:20
Reagent description	RGNT_DESC:C:25
Date of preparation	R_PREP_DT:D:9
Shelf life	R_SHELF_LF:N:8:0
Expiration date	R_EXP_DT:E:8
Expiration flag	R_EXP_FLG:E:7
Lot number of component #1	LOT_R1:C:20
Lot number of component #2	LOT_R2:C:20
Lot number of component #3	LOT_R3:C:20
Lot number of component #4	LOT_R4:C:20
Technician that prepared reagents	RGNT_TECH:C:5

Figure 4-4

The database in this example definition consists of 26 fields. A database can have up to 256 fields. The first field specified in Figure 4-3, Recno, is a field that is automatically created by @BASE. Do not input this name. Enter the first field name after Recno at the prompt. In this example, you would type ASSAY_TYPE. As stated in Chapter 2, if you are also using the field name in a spreadsheet template or macro program, the two names must be ***identical.*** For example, ASSAY_TYPE and ASSAYTYPE would be treated as two completely different entities by both Lotus and @BASE. If a macro program tried to access ASSAY_TYPE in a database file by calling it ASSAYTYPE, an error message would be issued. Also, field names must not contain any spaces and must begin with a letter. So, use caution when typing field names.

You do not have to use upper-case letters, @BASE will convert lower-case characters to upper-case automatically. Therefore, you may type either assay_type or ASSAY_TYPE. Press [RETURN] after you have typed the field name.

Specify the field type next. Field types can be character, numeric, logical, or date. The following review some of the specifics:

- ***Character*** fields store text and consist of any alphanumeric characters. If numbers are entered into the field, they are treated as characters and lose their ability to be manipulated mathematically.
- ***Numeric*** fields consist of numbers, decimal places, and +/- signs.
- ***Logical*** fields contain one character, **T** for true or **F** for false.
- ***Date*** fields are displayed in the MM/DD/YY format and are stored on disk in the dBASE III YYYYMMDD format. Dates are converted to the Lotus serial date

Standards Database (STANDARD.DBF)

Identification Fields

Record Number (From @BASE)	Recno:N:8:0
Lot number of entire std. set	LOT_STD:C:20

Information Fields

Name of standard set	STD_NAME:C:20
Standard set description	STD_DESC:C:25
Date of preparation	S_PREP_DT:D:9
Shelf life	S_SHELF_LF:N:8:0
Expiration date	S_EXP_DT:E:8
Expiration flag	S_EXP_FLG:E:7
Lot number of standard #0 (Blank)	LOT_S0:C:20
Concentration of standard #0	CONC_S0:N:10:3
Lot number of standard #1	LOT_S1:C:20
Concentration of standard #1	CONC_S1:N:10:3
Lot number of standard #2	LOT_S2:C:20
Concentration of standard #2	CONC_S2:N:10:3
Lot number of standard #3	LOT_S3:C:20
Concentration of standard #3	CONC_S3:N:10:3
Lot number of standard #4	LOT_S4:C:20
Concentration of standard #4	CONC_S4:N:10:3
Lot number of standard #5	LOT_S5:C:20
Concentration of standard #5	CONC_S5:N:10:3
Lot number of standard #6	LOT_S6:C:20
Concentration of standard #6	CONC_S6:N:10:3
Lot number of standard #7	LOT_S7:C:20
Concentration of standard #7	CONC_S7:N:10:3
Technician that prepared standards	STD_TECH:C:5

Figure 4-5

when they are used internally by @BASE.

Press [RETURN] to select Character for the field type of ASSAY_TYPE. Then type 10 and press [RETURN] to specify a length of 10. Each character field can be up to 254 characters long. With 256 fields per record, each record can have up to a total of 4,000 characters.

You have now entered the first field into the database definition. Repeat the procedure for the remaining fields, using the specifications given in Figure 4-3. When entering date fields, select *D*ate as the field type. @BASE will enter a length of 8 for you. If you select *N*umeric as the field type, you can specify numbers up to 19 characters long and up to 15 decimal places.

If you make a mistake in any of the field definitions, just continue. You cannot make changes during this phase of creating a database, but you will have a chance to edit the definition when you are done. When you have completed defining all of the fields,

Samples Database (SAMPLES.DBF)

Identification Fields

Record number (from @BASE)	Recno:N:8:0
Sample barcode number	BARCODE:N:8:0

Information Fields

Sample name/description	SAMP_NAME:C:20
Type of assay(s) requested	TYPE_REQ:C:32
Requester's Name	REQ_NAME:C:10
Date of request	REQ_DT:D:9
Requester's notes and comments	RQS_NTS:C:30
Department number	DEPT_NUM:N:3:0
Project code	PROJ_CDE:C:6
Date sample will be delivered	SMP_DEL_DT:D:9
Sample rack barcode number	RACK_BCD:N:8:0
Storage location	STOR_LOC:C:10
Sample Status	S_STATUS:C:4

Assay Results Database (RESULTS.DBF)

Identification Fields

Record number (from @BASE)	Recno:N:8:0
Sample barcode number	BARCODE:N:8:0
Type of assay performed	ASSAY_TYPE:C:10
Date assay performed	RUN_DT:N:5:0
Batch number	BATCH:N:3:0

Information Fields

Plate number	PLATE_NUM:N:8:0
Dilution Factor	DIL_FCTR:N:10:0
Raw data point #1	RAW_DATA1:N:8:4
Raw data point #2	RAW_DATA2:N:8:4
Results to be reported	RESULTS:N:10:3
Results Status	R_STATUS:C:4

Figure 4-6

press [ESC]ape to enter the edit mode.

In the edit mode, you can now use the arrow and [F2] keys to move to and edit any field names, types, lengths, or decimals. You can also add more field definitions or delete existing field definitions at the pointer position using the [INS] and [DEL] keys, respectively. When you are finished making changes, press [ESC]ape. To save the new database definition to disk, select *S*ave.

@BASE next opens the file for processing. An ***alias*** is an internal name that @BASE uses to distinguish between different open database files. @BASE can have up to a total of about 14 files open at once. Each time you open a database file, you indicate the alias to use with that file. The file name is the default alias, but you can assign a different alias whenever you open a file. An alias can be up to 12 characters long and can consist of alphanumeric characters and the underscore character ("_"). Select BATCH as the alias and press [RETURN].

To view the current status of the new database file, select ***F***ile and then ***S***tatus from the main @BASE menu. Then select BATCH and press [RETURN]. ***F***ile ***S***tatus displays current information about an open database file. ***F***ile ***S***tatus is a quick and easy way to obtain pertinent information relating to the current status of an open file. ***F***ile ***S***tatus displays the following information:

- Current alias.
- The .DBF file name (including drive and directory path).
- Date of the last update.
- Number of records in the file.
- Length of the records.
- Number of fields in each record.
- Active criteria for the currently selected file.
- Open index files.

The database definition is now complete and the file is ready to accept records. Press [RETURN] to advance to the main @BASE menu.

You should now take the time to create the other databases specified in Chapter 2. For your convenience, they are reproduced in Figures 4-4 through 4-6. Name the files REAGENTS, STANDARD, SAMPLES, and RESULTS to correspond to the names provided in the figures. ***When creating databases from the lists, do not create fields for Recno or items listed as "E" types.*** If you recall from Chapter 2, an "E" designation means that the field is not part of the database, but is calculated from data in the database using an equation.

CREATING A DATABASE DEFINITION USING *U*TILITY *F*ILE *T*RANSLATE

As noted at the beginning of this chapter, there are other ways to create database files and definitions. ***U***tility ***F***ile ***T***ranslate creates a new database file from existing Lotus records, using labels and data located in the spreadsheet. Both the database definition and data are automatically transferred to a new database file. This method of creating database definitions has certain advantages over the ***F***ile ***D***efine method described in the previous section. One advantage is that it saves typing if you have already created a template to be used with the database. Entering field names in a spreadsheet is also much easier than using the @BASE ***F***ile ***D***efinition utility. Perhaps the most important advantage is that you can be assured that field names in the spreadsheet and database are identical.

The following are some details that describe how @BASE creates a database using *U*tility *F*ile *T*ranslate:

- *U*tility *F*ile *T*ranslate automatically creates a temporary database definition from the spreadsheet. You can modify the temporary definition before any data are transferred to the file.
- Field names are taken from the first row of the table being translated.
- Field definitions of the new file are determined by examining the column width, cell contents, and format of the first data record.
  - Field types:
    - @BASE senses the type of data in the first row below the field names. If no data exist in the row, @BASE assigns the field(s) to be numeric.
    - If the first data cell is a label, string formula, or blank, @BASE assigns the field to be character.
    - If the first data cell is formatted for date display, @BASE assigns the field to be date.
  - Field lengths:
    - The length defaults to the column width, the length of the cell format, or the maximum field length in the database column for character fields.
  - Field decimals:
    - The number of decimals defaults to the display format number of decimals.
- @BASE recognizes the Lotus format for organizing database records. Each column on the spreadsheet comprises one field and each row represents one record. The first row of the database must contain field names.
- Database records can be located anywhere in the spreadsheet. A range is used to identify where Lotus data are located.

To illustrate *U*tility *F*ile *T*ranslate, create the template shown in Figure 4-7 by typing the field names specified. Next, use the /*R*ange *F*ormat *F*ixed (Lotus 1-2-3) or /*F*ormat *F*ixed (Symphony) command to set the format of cell C3 to zero decimal digits. The format of the cell will then correspond to an example field definition for PH_EXT. Next, use the /*W*orksheet *C*olumn *S*et-Width (Lotus 1-2-3) or /*W*idth *S*et (Symphony) command sequence to set the widths of columns B and C to 10 and 4, respectively. If you will note, the 3459 in cell C3 becomes a series of four asterisks (****) to signify that the cell is too narrow for the specified number to be displayed. There is a discrepancy between Lotus and @BASE that needs to be resolved when you use *U*tility *F*ile *T*ranslate. The column width needs to be set to 5 to display four-digit numbers, whereas the field definition for the database needs to be set to 4. (A similar discrepancy occurs between Lotus and @BASE for dates: Lotus needs column widths of 9 and @BASE needs field lengths of 8.) Therefore, you either need to proceed with the translation using a column

```
-------A--------B--------C------D--------E--------F---------G-------H--
1
2    Recno    REQ_NAME   PH_EXT
3              Mezei      3459
4
5
6
7
8
```

Figure 4-7

width of 4 and then correct the column width in the spreadsheet or you need to set the column width to 5 and edit the database definition. Either method is acceptable.

If you are using Symphony, you must attach the @BASEUTL.APP application program before you can access *U*tility *F*ile *T*ranslate. To do so, issue the [SERVICES] *A*pplication *A*ttach command sequence, highlight @BASEUTL.APP from the menu, and press [RETURN]. Press [ESC]ape until you return to the work environment. If you are using Lotus 1-2-3 with the @BASE Option Pac, you must attach the @BASEUTL.ADN add-in. To do so, issue the [ALT]-[F10] *A*ttach sequence, highlight @BASEUTL.ADN from the menu and press [RETURN]. Press [ESC]ape until you return to the work environment.

Access the @BASE menu. The command sequence to use is: *U*tility *F*ile *T*ranslate.

Specify the database range on the spreadsheet as B2..C3. When working with your template, be sure to include all field names and at least the first row of data in the table to be translated. ***Do not include Recno as a field name.*** As previously stated, it is important to have representative data in the first data row of a template. If there are no data in the range, @BASE will apply default values to all fields not containing data. Correspondingly, if incorrect column widths or non-representative data exist in the first row of cells, incorrect field types, lengths, etc., will be assigned.

Next, choose a name for the database destination file (e.g., PHONE). Type the name of the new file and press [RETURN]. ***If you choose the name of an existing file, @BASE will overwrite the existing file with the newly translated field definitions and spreadsheet data and previous definitions and data will be permanently lost.***

@BASE will respond with a field edit screen. This screen will give you a chance to modify field definitions before the data are translated. To modify the temporary definitions, select the desired field with the arrow keys and press [F2] to enter the edit mode. Press [ESC]ape when you are finished modifying the database definition. @BASE will then ask whether you want to cancel or save. Choosing *C*ancel aborts the translation and returns you to the @BASE menu. Choosing *S*ave saves the file definition and copies the worksheet data into the new database file.

CREATING A DATABASE DEFINITION USING *F*ILE *E*XTRACT

File ***E***xtract is a third convenient way to create database files and definitions. ***F***ile ***E***xtract creates a new database file from an existing database file. You can create a new database file that is identical to an existing database file or you can create a new file which is a subset of an existing database file.

If you choose this method,

- the file definition from the source file is automatically copied to the new file.
- data can be copied to the new database.
- criteria can be applied to data to be transferred so that only data that pass the criteria are copied to the new database.
- all, or a subset, of the fields in a file can be extracted.

If you have a database file created by Ashton Tate's dBASE III program, you can extract a file definition and data from the file using the following protocol. If not, you can use one of the database files created in the previous sections, because they are, in all respects identical to dBASE III-created files. To begin the process, issue the ***F***ile ***O***pen command from @BASE's main menu. Open the database file to be copied from. Then select ***F***ile ***E***xtract and choose the source file. Select an alias for the source database file and press [RETURN].

@BASE will respond by asking whether you want *A*ll of the fields extracted or *S*elected fields. If you choose *S*elect, @BASE displays a field selection screen. Use the arrow keys to highlight the desired fields to be extracted and press [RETURN]. After selecting all of the fields to be extracted, press [ESC]ape. The fields will appear in the new file in the order that they were selected from the source file.

Next, specify a file name and press [RETURN]. Choose an alias for the new database file. The new file is left open for @BASE processing, so that you can use the *C*opy menu choice to transfer the data from the specified fields in the source file to the new file.

EDITING DATABASE DEFINITIONS

After you have worked with a database file for awhile, you may find it necessary to make changes to the file's database definition. @BASE makes it easy to change field names, lengths, and types.

There are two commands on the @BASE menu that allow you to modify database definitions. If you are changing a field name, use ***F***ile ***F***ield-rename. If you are converting a field from one type to another or are changing the length of a field, use ***F***ile ***M***odify. Changing field names has no effect on the actual data in a file. However, modifying field types, lengths, and number of decimal digits might.

Field Type Conversion Table

Numeric ☞

- ***Character:*** Numbers are converted to alphanumeric characters.
- ***Logical:*** 0 is converted to F(alse) and 1 is converted to T(rue). Any other values are left blank.
- ***Date:*** Numbers in the range 1 to 73050 are interpreted as Lotus serial date numbers. 1 is converted to January 1, 1900 and 73050 is converted to December 31, 2099.

Character ☞

- ***Numeric:*** Converts if numeric (no alpha) characters exist in the field; otherwise, the field will be empty.
- ***Logical:*** (Not recommended.) Will convert only data beginning with a T or an F to a one-character logical field.
- ***Date:*** If the character field contains a valid date in the "MM/DD/YY" format, it is converted to the appropriate date. Any other value is left blank.

Figure 4-8

Changing Field Names

To change field names, use ***F***ile ***F***ield-rename. You cannot use ***F***ile ***M***odify to change field names. @BASE uses field names to identify fields. If you were to change a field name using ***F***ile ***M***odify, @BASE would not recognize the field as the same field. To @BASE, changing a field name with ***F***ile ***M***odify would be the same as deleting the original field and creating an entirely new one. Therefore, data in the field would be lost.

To rename a field, invoke @BASE and open the database file to be changed. Select ***F***ile ***O***pen from the menu. Next, select ***F***ile ***F***ield-rename and the name of the file you want to modify. You will be presented with a display of field names. Use the [UP] and [DOWN] arrow keys to highlight the field name to be modified. Press [F2] to enter the edit mode. Type the new field name and press [RETURN]. Press [ESC]ape when you are finished making changes. The changes are effective immediately.

When renaming fields, make sure that you review @DB functions in macro programs and update them to the new field names. If you have created index files for the database, you should also review the names in the index key expression for appropriateness. (See

Field Type Conversion Table

Logical ☞

- ***Character:*** Converts to a one-character field containing a T, F, or blank if the field is undefined.
- ***Numeric:*** T is conveted to 1, F is converted to 0.
- ***Date:*** Do not try to convert logical fields to date fields. All fields are left empty with this conversion.

Date ☞

- ***Numeric:*** Converts from the MM/DD/YY format to a number equal to that of the Lotus date serial number. Returns a positive integer between 1 (January 1, 1900) and 73050 (December 31, 2099). Numbers can be converted back to dates if they fall within the accepted serial number range.
- ***Character:*** Converts to a character string in the "MM/DD/YY" format. (October 18, 1990 is converted to 10/18/90.)
- ***Logical:*** All fields are left blank.

Figure 4-9

Chapter 5 for details on using ***I***ndex ***E***dit.)

Changing Field Types, Lengths, and Number of Decimal Digits

To change field types, lengths, and number of decimal digits, use the ***F***ile ***M***odify command sequence. When changing field types, @BASE automatically converts the data to the new format. Therefore, ***you must be very careful when changing field types to ensure that the conversion process does not cause damage to the data.*** If you change field types, be certain that the existing data in the field are in a form that can be converted to the new field types. If data types are incompatible, data for that field will be lost. For example, changing numeric fields to character fields usually presents no problem. However, changing character fields to numeric fields must be done only after the characters are converted to numbers. Otherwise the resulting fields will be left empty. Similarly, if you reduce character field lengths, the excess characters of any existing data in the fields are truncated from the end of the fields to match the new length. If you decrease field lengths of numeric fields, numbers that can no longer fit into the new field length are set equal to the null value (NA). Figures 4-8 and 4-9 outline the conversions. But, as a general rule, be very careful when designing your system and when making changes to it.

Because of the risk involved and the ease at which data can be lost as a result of changes, it is crucially important to avoid having to make changes to the database by planning your system prior to implementation. It is also important to prepare backup copies of your database files prior to making changes to database definitions. That way, if data are lost, you can always go back to previous versions and retrieve the lost data.

To make changes to a definition, invoke @BASE and open the database file to be changed. Do so by selecting ***F***ile ***O***pen from the menu. Next, select ***F***ile ***M***odify and the the name of file you want to modify. You will be presented with a display of fields. Use the [UP] and [DOWN] arrow keys to highlight the field definition to be modified. Press [F2] to enter the edit mode. Type the new field definition and press [RETURN]. When changing field lengths, only the field lengths of character and numeric fields can be changed because date and logical fields have set field lengths. Press [ESC]ape when you are finished making changes. To save the field definition changes and convert exiting data, choose ***S***ave from the update menu. To abort the changes, either choose ***C***ancel or press [ESC]ape.

Adding and Deleting Fields

Adding and deleting fields from existing database definitions can also be accomplished using the ***F***ile ***M***odify command. New fields can be inserted anywhere in the current database definition. To insert a field, open the file and access ***F***ile ***M***odify using the commands given in the previous subsection. Position the cursor where you want the new field to appear and press the [INS] key. A blank line will be inserted and you will be prompted to enter the new field definition. To delete a field, move the cursor to highlight the field to be deleted and press the [DEL] key. The field will disappear. If you decide that you do not want to add or delete fields, press [ESC]ape and then ***C***ancel. The changes you have indicated are not actually made until you ***S***ave them explicitly.

SUMMARY OF WHAT HAS BEEN ACCOMPLISHED

This chapter explained how to install and configure @BASE and the @BASE Option Pac and how to create database definitions.

When you build new databases, you must first create ***database definitions*** before you can enter data. For each field in a new database, a database definition contains specifications for field names, field types, field lengths, and numbers of decimal digits. There are three ways to create database definitions:

- Manually type the definition using @BASE ***F***ile ***D***efine.
- Translate the definition from a Lotus spreadsheet.

- Extract the definition from an existing database file.

Database definitions can be modified at any time. However, changes are risky and it is easy to lose data as a result of changes. Therefore, it is very important to plan your system prior to implementation. It is also important to prepare backup copies of your database files prior to changing definitions. That way, if data are lost, you can always go back to previous versions and retrieve lost data.

WHAT'S NEXT?

Now that you have created individual databases, Chapter 5 shows you how to create index files for them and how to join database files to create extended relational databases.

5

Creating Index Files and Joining Databases

It is now time to take the ***Consolidated Index list***, ***Data Flow Diagram***, ***Data Relationships For Joining Files list***, and ***Virtual Fields Specifications For Joining Files list*** (created as a result of Chapter 2) and convert the lists into index files and joined databases. This chapter explains how to use the @BASE Option Pac as an invaluable tool for

- creating and maintaining index files that are fully compatible with Ashton-Tate® dBASE III® index format.
- quickly locating any record in a database, no matter how many records are stored in the database.
- quickly sorting a database, without physically rearranging the records on disk.
- automatically preserving the proper sort-order of a database, even though records are added, deleted, or modified.
- joining two or more database files to create a single "virtual" database with true, extended relational properties.
- creating "computed" fields (whose values are calculated from other fields in the database), thereby conserving disk space.
- saving computed-field and joined-field set-up information in configuration files (called "View Files"), to allow immediate re-establishment of the infrastructure at system start-up.

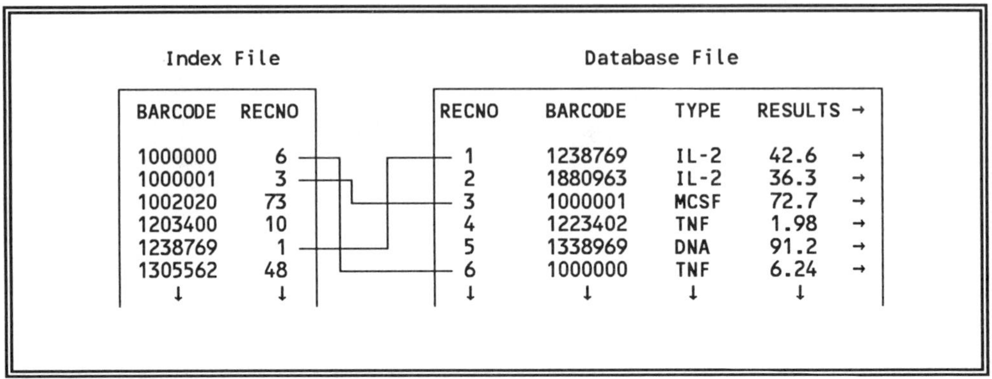

Figure 5-1

INDEX FILES AND KEYS

Perhaps the most important aspect of the @BASE Option Pac is its support for index files. This ability is extraordinarily powerful and adds enormous value to LIM systems.

To briefly review Chapter 2, database index files are special files that contain information about the location of records in a database. Using an index file and a ***key value***, you have quick access to records in a database. ***A key*** is a field in an index file that points to the locations of records in a database file (Figure 5-1). The index key

- is used in searches for records.
- is used in joining files to create extended relational databases.
- determines the order that the records in a data file are displayed.

You can have more than one index per database file. It is often useful to create several index files for a single database. You can also have ***compound indexes***, where the index key consists of more than one field. For example, in addition to a BARCODE index, a second index file containing a different set of keys (e.g., STATUS and REQ_DT), might be helpful to speed preparation of end-of-month reports. If more than one index is open for a database, the primary index is used for searches and also determines the order in which records are found. The first index opened for a particular database is the primary index until another index is selected as the primary index.

Once you have created index files for databases, the @BASE Option Pac automatically keeps the index files up-to-date. When records are modified, added, or deleted, index files are amended to maintain the proper sort-order.

INDEX KEY EXPRESSIONS

When you create an index file, you are actually creating an ***index key expression*** that is used by the index. The index key expression determines the order that records in the data file are displayed, as well as the key value used during record searches and joining of files to create extended relational databases. An index key expression is like a criteria expression in that it returns a computed value. However, unlike criteria expressions that return either true or false, index key expressions return either a numeric or string value. As an illustration, suppose a database had fields with the following field definitions:

- ASSAY_TYPE was defined as a character type.
- RUN_DT was defined as a date type.
- BATCH was defined as a numeric type with a field width of three and zero decimal digits.

The following example illustrates how these fields can be used in an index key expression:

```
UPPER(ASSAY_TYPE)+STR(YEAR(RUN_DT),4,0)+STR(MONTH(RUN_DT),2,0)
        +STR(DAY(RUN_DT),2,0)+STR(BATCH),3,0)
```

As you can see from the above equation, the index key expression that is created depends on the field definitions from a database. Because the above index key expression consists of more than one field, it is called a ***compound index key expression***.

There are two techniques that can be used to create index key expressions. Both methods are accessed through the @BASE menu system. The methods are:

- *I*ndex *P*rompt, which automatically generates an index key expression after you have selected the field names to be used in the expression.
- *I*ndex *I*nput, which allows you to manually type an expression of your own design.

In general, you probably will not have to manually type an index key expression for your system. You can use the *I*ndex *P*rompt command, select the key fields, and have the index key expression created automatically. However, you can create more complicated index key expressions using *I*ndex *I*nput. For example, the SAMPLES database contains a project code field (PROJ_CDE). If the last two characters in the field coded for a sub-project, a custom index key expression that focuses on the two end characters would be

```
RIGHT(PROJ_CDE,2)
```

The upper(), STR(), YEAR(), MONTH(), DAY(), and RIGHT() portions of the example equations are called @BASE ***operating functions***. The @BASE Option Pac supports a number of operating functions that can be used when creating index key expressions. @BASE operating functions are conceptually similar to Lotus string, numeric, date, logical, and database @functions. For a description of @BASE operating functions, see Appendix E. The remainder of this section explains some idiosyncracies of the @BASE operating functions that you need to be aware of.

- If an index key expression consists only of a single numeric field, the field will ***NOT*** be converted to a character string prior to use and searches of the index file will be performed numerically.
- If numeric fields are included in a compound index key expression created using the ***I***ndex ***P***rompt command sequence, @BASE automatically converts the numeric values to characters using the STR() operating function. The index is then treated as a character string index. For example, the ***I***ndex ***P***rompt command was used to create the compound index key expression at the beginning of this section and the index key expression for the BATCH field (defined as three numeric digits with no decimal digits) contained the following:

```
STR(BATCH,3,0)
```

- If you include numeric fields in a compound index key expression created using ***I***ndex ***I***nput, you must manually input the STR() function to convert numeric values to characters.
- Some operations are not allowed in index key expressions. For example, you cannot use

    - logical fields.
    - joined or computed fields.
    - cell references.
    - the functions "Recno", "Del", "Instr", "If", "Qtr", and "Today".
    - the values of ".NA.", ".T.", and ".F.".

- The @BASE setting that ignores case has no direct influence on index order. However, ***I***gnore-case determines how index key expressions are designed. For example, if @BASE was set to be case-sensitive and ***I***ndex ***P***rompt was used to create an index key expression on ASSAY_TYPE, "DNA" and "dna" would be considered differently and would not appear consecutively in a display of the file. However, if @BASE was set to ignore case, the ***I***ndex ***P***rompt command would automatically place an UPPER() function before character fields in the index key expression. For example, the index key would contain

```
UPPER(ASSAY_TYPE)
```

- When using the UPPER(ASSAY_TYPE) index key expression, the index entries for "DNA" and "dna" would be considered the same and the two records would appear consecutively in the file. However, if you create an index key expression using *I*ndex *I*nput, you must manually input the UPPER() function if you want the index to ignore case.

- Index key expressions and the resulting index keys (values returned by the index expression) cannot exceed 100 characters. If you exceed this limit, @BASE will issue an "Index expression too long" message, indicating that the index key expression, or the resulting string, is too long. This limit appears to be quite large. However, it is not and you may receive the message, especially if you are using compound key expressions and date fields. If you receive the message, try decreasing the length of field names or consider changing field types. Note the length of the index key expression at the beginning of this section. Three operating functions were required to convert dates to numeric and three STR() operating functions were required to transform the converted numbers to strings. These six operating functions add up to a substantial number of characters and can quickly add up to enough characters to exceed the limit. However, had the RUN_DT field been defined as numeric (with a field length of five and zero decimal digits) and a Lotus @DATE @function been used to convert the date to a serial number prior to transferring the date to the database, the index key expression would have been much shorter.
- It is important to keep index key expressions as small and compact as possible. Remember that an expression is re-evaluated for each record in the index file. If there are thousands of records in the file, a substantial amount of time may be required. Avoid compound index key expressions involving dates. As explained above, dates require six operating functions to be executed before the dates are converted to usable form. If you build your own user interface, converting dates to numbers with the Lotus @DATE @function prior to storage is highly recommended. If you use the @BASE or dBASE® III user interface or dBASE III, then the choice is a little more complicated. Dates will not be displayed in date format unless the field type is specified to be date. Therefore, user input and review would be impaired because the user would have to work with number equivalents of dates.

To focus on the importance of the above discussion on index design, consider the RUN_DT definition in the BATCH and RESULTS databases. Note that RUN_DT was defined as numeric in both databases. The above discussion on dates in index key expressions should clarify the reasons for using numeric definitions instead of date definitions.

The next section explains how to create index files that can be used for sorting, searching, and joining databases.

CREATING INDEX FILES

In Chapter 2, a ***Consolidated Index list*** was designed. The list specified field names for a series of index files. For your convenience, the list has been reproduced in Figure 5-2. The index files in Figure 5-2 collectively service all of the search, sort, and join functions

Consolidated Index List

DATABASE FILE	INDEX FILE	INDEX FIELD #1	INDEX FIELD #2	INDEX FIELD #3
SAMPLES	SAMPL_BC	BARCODE	-	-
RESULTS	RES_BC	BARCODE	-	-
RESULTS	RES_TRIO	ASSAY_TYPE	RUN_DT	BATCH
BATCH	BCH_TRIO	ASSAY_TYPE	RUN_DT	BATCH
STANDARD	STDS	LOT_STD	-	-
REAGENTS	RGNTS	LOT_RGNT	-	-

Figure 5-2

of the example LIM system. As stated in Chapter 2, the list has been optimized to eliminate redundant fields. After you have created a ***Consolidated Index list*** for your system, creating index files is trivial.

To create the first index file in Figure 5-2 (SAMPL_BC), access the @BASE menu. The keystrokes to access the @BASE main menu depend on the Release of your software:

- Lotus 1-2-3, 2.0, 2.01, 2.2: [SHIFT]-[F9].
- Symphony 1.2, 2.0: [F10], @Base
- Symphony 2.2: [SHIFT]-[F10], ***B***ase, [F10]

After you have issued the appropriate keystrokes, you will be presented with @BASE's main menu. From the main menu, select ***F***ile ***O***pen. Then highlight SAMPLES and press [RETURN]. Press [RETURN] again to accept SAMPLES as the alias.

If, when using this system, you receive a "Cannot open file," "Locked file," or "Cannot create file" error message, check your CONFIG.SYS file for a FILES command to ensure that the command is present and specifies at least 20 files. If FILES=20, then a maximum of about 14 files are available because Lotus and @BASE normally maintain up to about 6 open files. Because your system may utilize up to 14 open files, use your word processor to change the FILES statement to at least 20 and re-boot your computer. (The maximum number of files is 255. However, DOS allows no more than 20 files to each software application. So, increasing the number to more than 20 files will not increase the number of files available to Lotus unless you have terminate-stay resident (TSR) files running that also have open files. If your system exceeds the 14 open file limit, then you will need to review the system and try to combine database files or close open index or database files to decrease the number of open files to an acceptable limit.)

Index ***P***rompt builds an index file by letting you select fields from an open database file. From the main menu, select ***O***ptions ***I***ndex (***F***ile) ***P***rompt.

(NOTE: For Symphony, Release 2.2, the sequence for most of the commands listed in this chapter is ***O***ptions ***I***ndex ***F***ile, followed by the specified index menu choice. All other Releases do not contain ***F***ile in the sequence. For this reason, the extra ***F***ile command will be placed in parentheses for the remainder of this chapter to signify that it is for Symphony 2.2 use only.)

Highlight SAMPLES as the database file and press [RETURN]. Next, type SAMPL_BC as the index file name and press [RETURN]. You will be presented with a display of all of the field names in the SAMPLES.DBF database file. Use the arrow keys to move the cursor to the BARCODE field name. Press [RETURN] to select the field. BARCODE will be placed in the index key expression. To end the selection process, press [ESC]ape and then choose ***E***xecute from the menu. The SAMPLES.NDX index file will be saved to disk.

Repeat the above process to create all of the example index files specified in Figure 5-2. To select more than one field for an index key expression, move the cursor to the most important field name for the key and press [RETURN]. Then highlight the second field name, press [RETURN], and so on. @BASE builds the index key expression based on the order of selection. Therefore, if the ASSAY_TYPE, RUN_DT, BATCH trio were being specified, you would highlight ASSAY_TYPE first and BATCH last.

If you make a mistake and want to un-select an item, highlight the item and press [RETURN] again. [RETURN] toggles the selection of field names to be used in an index key expression.

If you receive an "Index expression too long" message after you press [ESC]ape, amend your fields using the tips provided in the previous section. If you decide to change a field name or type, see Chapter 4 for instructions on how to make the changes.

If you would like to review the index key expression for one of the index files, use the ***I***ndex ***E***dit command sequence. The primary purpose of ***I***ndex ***E***dit is to modify existing index key expressions and rebuild index files. However, ***I***ndex ***E***dit can be used to view an index key expression. An index file must be open to use ***I***ndex ***E***dit. If the index file is not open, use ***O***ptions ***I***ndex (***F***ile) ***O***pen to open it. When prompted for the database file name, highlight the file name and press [RETURN]. When prompted for the index file name, use caution when selecting the name from the menu to ensure that the index is one that corresponds to the selected database file. @BASE does not selectively exclude index files. The index file menu contains a listing of ***all*** of the index files in the subdirectory, whether they are appropriate to the selected database or not. If you receive a "File not open" or "Index not open" error message when working with index files, the chances are high that you opened an index file that was not created for the database that you selected.

After opening a file, select ***O***ptions ***I***ndex (***F***ile) ***E***dit from the menu, highlight the indexed database file and press [RETURN]. Select an index file from the menu and press [RETURN]. Then, use the left and right arrow keys to move the cursor down the index key expression. When an index key expression is longer than the width of the screen display, ***characters will be hidden in the middle of the expression***. Thus, the edit mode provided by @BASE is slightly different than the edit mode for spreadsheet cell contents. Therefore, you must use the left arrow key to expose hidden characters.

SPECIFYING THE PRIMARY INDEX FILE

The RESULTS.DBF database file in Figure 5-2 has two index files open when the LIM system is used (RES_BC and RES_TRIO). The ***primary index file*** is used for searches and also determines record order when using commands such as ***B***rowse and ***E***dit, and when using @BASE @DB @functions. For this reason, it is important to specifically state which index file is the primary index file.

The ***I***ndex ***S***elect-primary command sets the primary index for an open database. To set the primary index for the SAMPLES.DBF file, open the file and select ***O***ptions ***I***ndex ***S***elect-primary. Highlight SAMPLES and press [RETURN]. Select the SAMPL_BC index file as primary.

EDITING INDEX DEFINITIONS

Unlike modifying database definitions, modifying index definitions has no effect on the data in a database. There are two ways to change the index key expression controlling the indexing process. The two methods are:

- Re-create the index key expression from the start using the ***I***ndex (***F***ile) ***P***rompt command.
- Modify the index key expression using the ***I***ndex (***F***ile) ***E***dit command.

Both methods have their merits. As previously shown, the ***I***ndex ***P***rompt method is very fast, convenient, and produces reliable results. The ***I***ndex ***E***dit method requires more understanding of the use of operating functions, but can be used to create more powerful index key expressions. ***I***ndex ***E***dit works identically to ***I***ndex ***I***nput. Instructions for how to use ***I***ndex ***P***rompt, ***I***ndex ***I***nput, and ***I***ndex ***E***dit are provided in the previous section.

JOINING DATABASE FILES

Index files give you the ability to join multiple files to create extended relational databases. If you recall from Chapter 2, an extended relational database is a group of two or more files that work together in a special way. As illustrated in Figure 5-3, information that might otherwise be stored in a single large file with many fields is instead organized in several files, each with fewer fields. The files are ***joined*** together (***linked***) so that information can be shared among the files. In a sense, one data file can look up information in one or more other files. When individual database files are linked together to extend the ready availability of related information, the resulting system appears to be a single, massive file, containing many fields. This system is often referred to as a

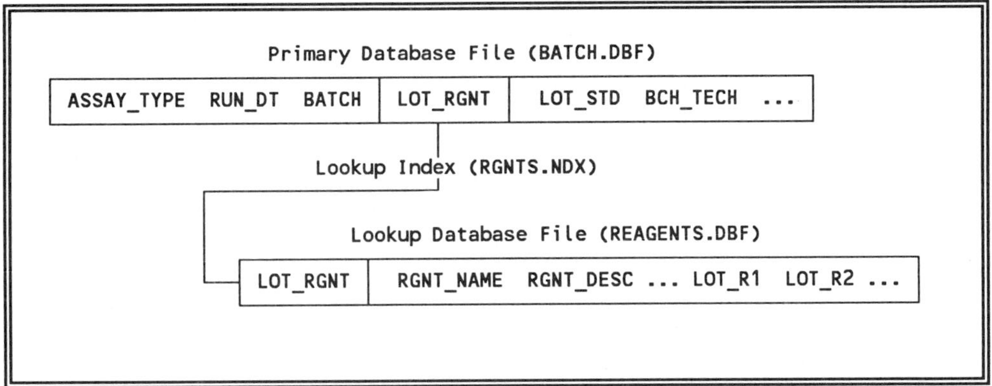

Figure 5-3

virtual database or an ***extended relational database***.

A virtual database consists of a primary file and one or more lookup files. The information that is unique to each record is stored in a file, called the ***primary file***. Information that is shared by many primary file records is stored as a single record in a separate file, called the ***lookup file***.

To join files, there must be at least one field that is common to both files being joined (e.g., LOT_RGNT in Figure 5-3). Further, when databases are joined, the common field must have the ***exact*** same name in both the primary and lookup files. ***The lookup file must be indexed on the field that is common to both lookup files.*** The databases are joined, using the lookup index file as the active link between the two database files. The index file link provides extended lookup capabilities.

The following is a list of some important details that you need to remember when joining files:

- The *J*oin command combines two database files based on an index file for the lookup file and one or more fields that are common to both files.
- There must be an open index file on the common field of the lookup file before the join can be performed.
- Each record in the lookup file contains information that can be shared with many records in the primary file.
- The fields selected for viewing as a result of the join (also known as ***virtual fields***) are logically added to the primary file. That is, the logically added virtual fields disappear when the session is ended.
- Virtual fields can be displayed and used in calculations, but cannot be edited or used in an index key expression.
- When you close one of the files or exit @BASE, the join parameters (primary file, lookup file, lookup file index, and virtual fields) are all lost. The join parameters can be saved to a view file and retrieved later. (See below.)

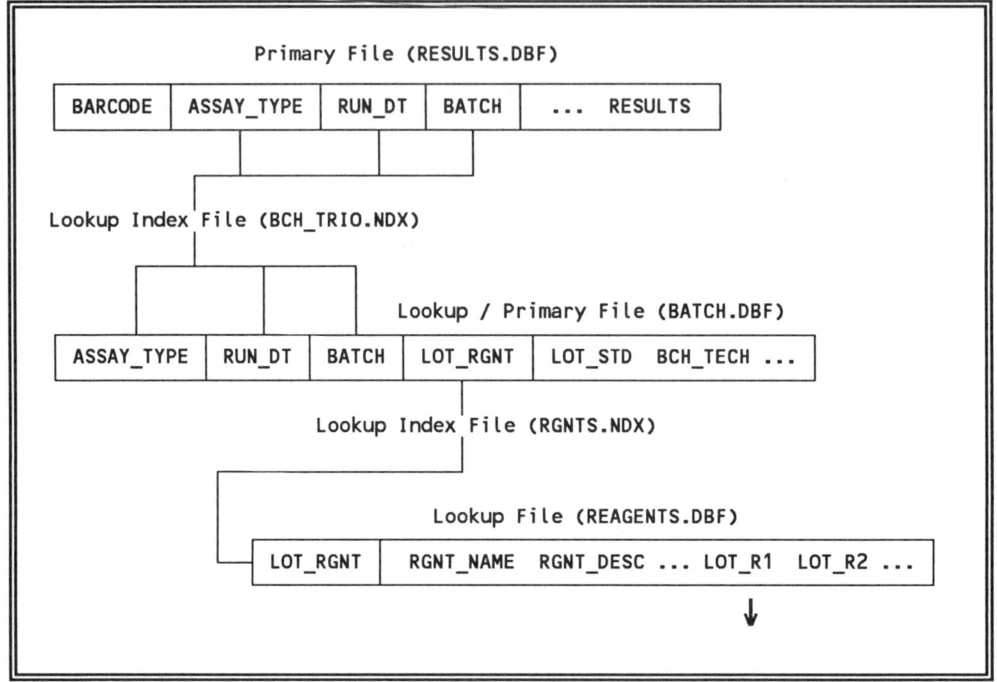

Figure 5-4

Files can be joined through several layers (as illustrated in Figures 5-4 and 5-5) to create a hierarchy. The hierarchy, in turn, makes field data available to each sequentially higher primary database file in the series. For example, in Figure 5-4 BATCH.DBF serves as both a lookup file (for RESULTS.DBF) and a primary file (for REAGENTS.DBF). The example system illustrates how specific information about a set of reagents used during an assay run of a sample can be made available to the sample's record.

The order of creation of links through a hierarchy of primary and lookup files is very important. If, in the example, RESULTS.DBF and BATCH.DBF had been joined prior to BATCH.DBF and REAGENTS.DBF, the BATCH.DBF virtual fields would not have been available to RESULTS.DBF when the fields were being selected.

The creation of extended relational databases consists of three tasks:

- Creating index files for the ***lookup*** database files.
- Joining the database files using the lookup index files.
- Saving the join parameters in a view file for future system start-up.

In the previous section, you created index files that could be used to join the files specified in Figure 5-6. The next subsection explains how to join two database files and

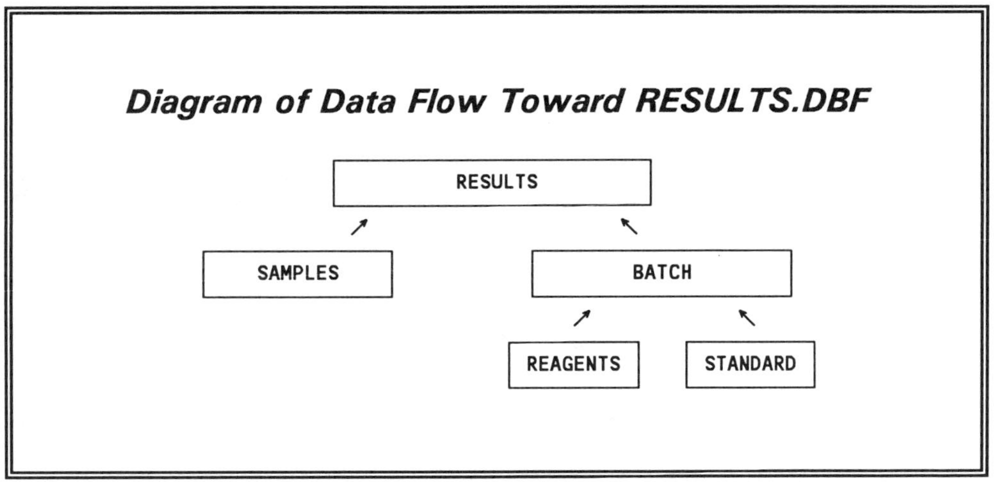

Figure 5-5

then add a third to the series. The final subsection explains how to create view files to save the configuration of a system.

Joining Databases

As just stated, the order in which links are created when joining files is very important. The arrangement of files to be joined in Figure 5-6 is inversely related to the hierarchy of the system. The flow diagram in Figure 5-5 was developed in Chapter 2 and illustrates the relationships of the files in the series. As you can see from Figure 5-5, the SAMPLES.DBF → RESULTS.DBF and BATCH.DBF → RESULTS.DBF links are higher in the flow diagram than the following links:

- REAGENTS.DBF → BATCH.DBF
- STANDARD.DBF → BATCH.DBF

Positioning affects the ability of a primary file to obtain information from lookup files in a flow diagram. The hierarchy of this flow is reflected in the creation order for the Data Relationships for Joining Files list in Figure 5-6. To preserve the correct hierarchy, start at the top of the Data Relationships for Joining Files list and work your way down the list when creating links. In this case, the first link to create is the one that uses BATCH.DBF as a primary file and REAGENTS.DBF as a lookup file. The index file to the lookup file is RGNTS.NDX and was created in a previous section of this chapter.

To create the links, access the @BASE menu system. An index file must be open to use ***I***ndex ***E***dit. If the index file is not open, use ***O***ptions ***I***ndex (***F***ile) ***O***pen to open it. Before an index file can be opened, its associated database file must first be opened. For future convenience, use ***F***ile ***O***pen and ***O***ption ***I***ndex (***F***ile) ***O***pen to open all of the

Data Relationships for Joining Files List

HIERARCHY	PRIMARY DBF	LOOKUP DBF	LOOKUP NDX	JOIN FIELD #1	JOIN FIELD #2	JOIN FIELD #3
LOWEST	BATCH	REAGENTS	RGNTS	LOT_RGNT	-	-
	BATCH	STANDARD	STDS	LOT_STD	-	-
	RESULTS	SAMPLES	SAMPL_BC	BARCODE	-	-
HIGHEST	RESULTS	BATCH	BCH_TRIO	ASSAY_TYPE	RUN_DT	BATCH

Figure 5-6

database and index files on the list. When prompted for the database file name, highlight the file name and press [RETURN]. When prompted for the index file name, use caution when selecting the name from the menu to ensure that the index is one that corresponds to the selected database file. @BASE does not selectively exclude index files. The index file menu contains a listing of ***all*** of the index files in the subdirectory, whether they are appropriate to the selected database or not. If you receive a "File not open" or "Index not open" error message when working with index files, the chances are high that you opened an index file that was not created for the database being used.

Next, select ***O***ptions ***L***ink from the main menu. You will be prompted through the next sequence of steps:

- Select the primary file (BATCH.DBF) from the menu and press [RETURN].
- Select the lookup file (REAGENTS.DBF) from the menu and press [RETURN].
- Select the index file (RGNTS.NDX) from the menu and press [RETURN].
- You will be presented with a list of all of the fields in REAGENTS.DBF. Select the fields (e.g., R_PREP_DT and R_SHELF_LF) that you want to appear when viewing the primary file (BATCH.DBF) by highlighting the fields and pressing [RETURN]. Select the fields in the order that you want them to appear in the primary file. Do not use [RETURN] to toggle selected items and do not select the common field (i.e., the indexed field). @BASE does not allow duplicate field names in a file. The fields you select will become virtual fields and will be logically added to the primary file.

Repeat this process for each item on the list in Figure 5-6. Start with the next item on the list and work your way down. Figure 5-7 provides the virtual fields to select, in order of selection. Note that all field names in Figure 5-7 are unique. @BASE does not allow

Virtual Fields Specifications for Joining Files

PRIMARY DBF	LOOKUP DBF	FIELD #1	FIELD #2	FIELD #3	FIELD #4	FIELD #5
BATCH	REAGENTS	R_PREP_DT	R_SHELF_LF	-	-	-
BATCH	STANDARD	S_PREP_DT	S_SHELF_LF	{CONC_S0	THROUGH	CONC_S7}
RESULTS	SAMPLES	SAMP_NAME	TYPE_REQ	REQ_NAME	DEPT_NUM	PROJ_CDE
		REQ_DT	-	-	-	-
RESULTS	BATCH	{ALL FIELDS, BOTH VIRTUAL AND REAL}				

Figure 5-7

duplicate field names, whether the field names are within the database or are becoming part of the database as a result of another database being joined to it. Indexing and joining database files often reveal redundancy and inconsistencies in field names and types. These problems are especially apparent when joining files. Although two or more database files can use identical field names, if the field names become virtual fields of the same file, @BASE will issue an error message. For this reason, it is good policy to review the lists of fields for each database file created in Chapter 4 to ensure that they are correct and make modifications if needed.

To see the effect of the linking process, issue the ***D***ata ***B***rowse command. Choose RESULTS as the database file and press [RETURN]. Use the [RIGHT] arrow key to move down the list of field names. You should notice a change in the field names from upper-case to lower-case. @BASE uses lower-case field names to distinguish between virtual and true fields.

Removing Virtual Fields and Linked Databases

To remove virtual fields from an extended relational database, access ***F***ile ***M***odify and specify the name of the primary database file. When presented with the list of field names, use the [DEL] key to remove field names from the list. Press [ESC]ape when you are done and select ***E***xecute from the menu.

To delete a link between databases, delete all of the virtual fields in the primary database that pertain to the lookup file(s) being unlinked using ***F***ile ***M***odify. Then, close the lookup file(s). If you have used a view file to create the link, issue an Options ***V***iew ***S***ave command to re-save the view file. (See next subsection.)

Creating View Files

Creating all of the links in the previous subsection can be quite tedious. Unfortunately, all virtual relationships are lost when files are closed or @BASE is exited. Because joined files are virtual relationships, the work that you just completed in the previous section would be lost at the end of this session. However, if you create a view file for each open database file before closing the files and leaving @BASE, you can easily retrieve the view files the next time you start @BASE and retain the joined database definitions.

A ***view file*** is a way of saving the current relational configuration of a system in a file so that the setup information can be easily retrieved at a later time. When a view file is saved, the join definitions presently in effect for the specified primary file are recorded on disk. Later, when the view file is retrieved, the recorded definitions for the file are immediately re-established and the system is set up automatically. A view file, in effect, takes a "snapshot" of the system and saves the image on disk. The view file is created automatically whenever a *V*iew *S*ave command sequence is performed.

A view file consists of a series of statements that are interpreted as commands. The statements are executed in order. When a view file executes, it

- opens the primary and all listed lookup database files and assigns database aliases.
- opens all listed index files and assigns index aliases.
- creates virtual fields to form active links between the primary and lookup file(s).

Only lookup files directly linked to a primary file are opened. For this reason, a view file for RESULTS will open SAMPLES and BATCH, but not REAGENTS and STANDARD. For this reason, a view file for BATCH also needs to be created.

To create a view file for BATCH.DBF:

- Make sure all database and index files have been opened and joins have been completed.
- Delete all virtual fields and close all index files that you do not want to include in the view file.
- Select *V*iew *S*ave from the menu.
- Select the database file (BATCH.DBF) from the menu and press [RETURN]. (Make sure that you are specifying the primary file and not the lookup file.)
- Name the view file. Typically you would use the name of the primary file. In this case, specify BATCH and press [RETURN].

Repeat the process for the RESULTS database. The next time you begin a session you must execute the view files in reverse order of the hierarchy. That is, you must execute the view files in order from lowest member to highest or the virtual fields will not pre-exist for primary files higher on the hierarchy list and the system will fail. The following

Computed Field Operators and Operating Functions

RESULT TYPE	MATHEMATICAL OPERATORS	OPERATING FUNCTIONS
NUMERIC	+, -, *, /, ^	INT, ROUND, ABS, VAL, DAY, MONTH, YEAR, IF, RECNO, INSTR, LEN
DATE	+, -	CTOD, IF, TODAY
LOGICAL	.AND., .OR.	DEL, IF
CHARACTER	+ (STRING CONCATENATION)	LEFT, LOWER, LTRIM, RIGHT, RTRIM, STR, SUBSTR, TRIM, UPPER, DTOC

Figure 5-8

is the order of execution for this example system:

- BATCH.VU first.
- RESULTS.VU second.

To execute a view file, simply select *View Open* from the menu, select the name of the view file, and press [RETURN]. To display the virtual field definitions in an existing view file, select *View List* from the menu, select the name of the view file you want to review, and press [RETURN]. A list of the origins of the virtual fields will be displayed. To exit the display, press [RETURN]. If the list is longer than the screen can display, you can use the [DOWN] arrow key to scroll the screen to view the hidden data.

There is one other detail about view files to be aware of. With Symphony 2.2, there was a change in file extension (from .VU to .VU2) and file format. Because of the change in file format, you must convert .VU files if you upgrade from previous Lotus 1-2-3 or Symphony Releases to Symphony 2.2. A file on the @BASE Option Pac diskette (called "VUVU2") will perform the conversion for you.

CREATING COMPUTED FIELDS

A ***computed field*** is another category of virtual field. Computed fields are fields whose values are calculated from other fields in a database. For example, expiration dates can be calculated using date of preparation and shelf life fields. Computed fields are logically part of databases but, unlike normal fields, their values are not stored on disk. There is no practical reason to consume disk space for field values that can be derived from other fields.

Because computed fields are virtual fields, temporary, and last only as long as a file is open, they are guaranteed to be current. However, computed fields need to be created at system start-up (either manually or using view files).

The following is a list of some important details that you need to know when using computed fields. Computed fields

- are equations consisting of field names, functions, and constants.
- can contain cell addresses and range names from the spreadsheet.
- can contain virtual fields.
- are limited to 240 characters.
- can be character, numeric, logical, or date fields.
- are not physically attached to the target file; but consist of temporary field definitions, and the contents of these fields are computed when they are accessed.
- return virtual values that cannot be edited.

Computed fields contain the results of calculations based on values in other fields of the same record. ***Computed field expressions*** are formulae that are evaluated to return a single value. The value returned can be a numeric, logical, date, or character string. A number of operating functions or operators may be used in a computed field expression. However, functions and operators must be consistent with the data types being evaluated by the expression. Expressions can be built using fields in the database, constants, and contents of spreadsheet cells. Figure 5-8 lists some of the operators and operating functions that can be used in computed field expressions. For a more thorough explanation of the operating functions in Figure 5-8, see Appendix E.

If you recall from Chapter 2, there were several fields in the database files that we designated as "E" types. "E" type fields can be calculated using computed fields. To illustrate computed fields, consider a common use in which expiration dates are determined by calculation from date of preparation and shelf life. To create a computed field to calculate the R_EXP_DT (reagent expiration date) field of REAGENTS.DBF using the R_PREP_DT (reagent preparation date) and R_SHELF_LF (reagent shelf life) fields, open the REAGENTS database and select ***O***ptions ***C***omputed_field ***C***reate from the menu. When prompted for the database file name, select REAGENTS.DBF and press [RETURN]. When prompted for the field name, type R_EXP_DT and press [RETURN].

A menu will appear to allow you to choose the field type. Choose the field type with caution. The following rules apply:

- If a computed field consists of character fields and expressions, select character.
- If a computed field consists of numeric fields and expressions, select numeric.
- If a computed field consists of a combination of types, be sure to include the proper conversion functions.

Because this example displays an expiration date, choose ***D****ate*. Enter the following computed field definition and press [RETURN]:

```
R_PREP_DT+R_SHELF_LF
```

Repeat the process to create a computed field for the expiration flag. This time, the name of the field should be R_EXP_FLG, the field type should be ***L****ogic*, and the expression should be:

```
IF(TODAY()<R_EXP_DT,.T.,.F.)
```

To review the results of these actions, issue a ***D****ata* ***E****dit* command. When prompted for the database file, choose REAGENTS from the menu and press [RETURN]. When presented with a display of the fields, use the arrow keys to highlight the DATE_PREP field, press the [F2] key to enter the edit mode, type in some representative data, press [RETURN], and examine the changes that take place in the R_EXP_DT and R_EXP_FLG fields. Repeat the sequence to modify the R_SHELF_LF field and examine the changes.

To edit computed field definitions, use the ***O****ptions* ***C****omputed_field* ***M****odify* command. When presented with a listing of the fields, highlight the computed field that you want to modify and press [RETURN]. You can only change computed field definitions using the ***C****omputed_field* ***M****odify* command. You will be prompted for the changes. Press [RETURN] when you have completed the changes on the computed field. Press [ESC]ape to return to the main menu.

When you are finished, use the ***V****iew* ***S****ave* command to save the computed virtual field definitions to disk. Name the file REAGENTS.

If you join files and use link computed fields, you must create and use a view file(s) for the lookup file(s) containing the computed fields before executing the view file for the primary database. Otherwise, the computed field names would not exist and an error message would be issued. For example, if you want r_exp_dt available as a virtual field to BATCH.DBF, you must execute REAGENTS.VU before BATCH.VU.

For practice, repeat the above process to create computed fields for the other fields specified as "E" in Figures 2-10 and 2-11 and save the view files.

SUMMARY OF WHAT HAS BEEN ACCOMPLISHED

This chapter explained how to put some real power into your database system. The chapter showed how to index and join database files, create view files, and create computed fields.

The following are some definitions that were provided and are important to remember:

- An ***index key expression*** is an equation that determines the order of key values in an index file. An index key expression also specifies how the data in key fields will be used to determine the locations of records in the associated database file.

- ***Compound index key expressions*** are index key expressions containing more than one field.
- ***Operating functions*** are @BASE functions, similar in concept to Lotus @functions, that are used in index key expressions and computed field equations.
- A ***view file*** is a way of saving the current relational and computed field configuration of a system in a file so that the setup information can be easily retrieved at a later time.
- A ***computed field expression*** is an equation that is evaluated to return a single value. The value returned can be a numeric, logical, date, or character string.

Indexing and joining database files often reveal redundancies and inconsistencies in field names and types. These problems are especially apparent when joining files. Although two or more database files can use identical field names, if the field names become virtual fields of the same file, @BASE will issue an error message. For this reason, it is good policy to review the lists of fields for each database file to ensure that they are correct and make modifications if needed.

WHAT'S NEXT?

The next chapter provides a concise explanation of the Lotus Command Language and provides macro programming fundamentals. An understanding of these fundamentals is a prerequisite to creating programs to manage data flow throughout user interfaces.

6

Macro Programming Fundamentals

A set of instructions in a format and language that Lotus® can understand is called a Lotus ***macro*** program. The language that is used is called the ***Lotus Command Language*** (LCL) and the instructions are called ***macro commands***. An overview of the six broad categories of Lotus macro commands was provided in Chapter 1. The macro commands of the Lotus Command Language relevant to this book are given in Appendix C.

Macro programs are invaluable tools. In addition to exercising your creativity, macro programs can be used to place values in cells rather than assigning the task to active formulae. There are many situations where I use macro programs. The following are some situations when I use macros and the motivation of my preference for using the macro approach:

- As an alternative to active formulae in cells. The more formulae that you have in cells, the slower a spreadsheet will recalculate. If you have been creating large templates, you have undoubtedly noticed that the recalculation speed of the spreadsheet became progressively slower as more cells were added. The length of time required for recalculation would be especially evident if you had a PC or XT personal computer and/or had been using @BASE @DB @functions in your spreadsheets. As noted in Chapter 3, each time Lotus performs a spreadsheet recalculation, @BASE recalculates all @DB @functions in the spreadsheet. Every time an @DB @function is evaluated, @BASE starts at the beginning of a database file and reads each record until it locates the information it is searching for. This process is typically slow. If a spreadsheet contained many @DB functions, and if a database file contained thousands of records, the spreadsheet could take a very long time to recalculate. ***Therefore, you should avoid at all costs the use of @DB @functions in spreadsheet cells.***

You could diminish the problem by issuing the /**R**ange **V**alues command to convert formulae to their values, but then you would have to re-type the formulae into the template each time you wanted to re-use the template with new data. With macro programs, the formulae permanently reside in the spreadsheet, are reusable, and do not affect the recalculation speed of the template. However, because values in cells do not recalculate when they have been placed there by a macro program, you will need to decide which cells of the spreadsheet need to have active formulae and which cells need to have their values placed by macro programs.

- To increase the accuracy of displayed @DB values. As stated earlier, @DB @function recalculation is under the control of the spreadsheet. The values displayed are therefore dependent on the recalculation settings of the spreadsheet (e.g., minimum recalc), rules that force spreadsheet recalculation, and whether or not recent user activity has initiated a recalculation. Therefore, to increase the accuracy of the displayed values of @DB @functions, it is better to explicitly execute @functions using macro programs than to have @functions in spreadsheet cells.
- To decrease the chance of user mistakes. To use the features of a LIM system, a user would be required to issue commands to activate Lotus and @BASE features. These command sequences would obligate the user to properly issue commands, define named ranges, and specify appropriate ranges when prompted for them. These details could lead to incorrect actions and, in turn, could lead to misinterpretation of experimental data. If results were automatically transferred to a template by a macro program, the results could be reviewed with increased confidence. Another degree of safety could be achieved when a macro program places the results of analyses directly into a template as data are being acquired. The focus of ***Laboratory Lotus®: A Complete Guide To Instrument Interfacing*** and ***Practical Spreadsheet Statistics and Curve Fitting for Scientists and Engineers*** was to perform data reduction as an integral part of data generation. If you have integrated your instruments and spreadsheets using the concepts of those books, then you should think about how to add database management capabilities to the existing programs.
- To improve the "user friendliness" of a spreadsheet. User friendly spreadsheets (spreadsheets that make the user's tasks as easy to discern and accomplish as possible) allow users with little specialized knowledge to use your spreadsheets. As a general rule, the easier it is for a user to learn how to operate a program, the less training that you will need to provide. Spreadsheets can be made user friendly by including programs that have menus, prompts for information, graphics, status messages, and help messages. This type of system will make it much less frustrating for the users of your spreadsheet.
- To improve the user interface. A ***user interface*** is the means by which a user

interacts with the LIM system. As noted in Chapter 3, the user inputs, edits, and views data via the user interface. Preventing mistakes at this level is very important because it is the one element that you cannot fully control through programming. Therefore, it is crucial that you create optimal designs for user interfaces. Unfortunately, the @BASE interface is limited because data are presented either in columns or rows. These formats tend to make the display lack organization in terms of related information. Furthermore, @BASE uses field name acronyms in the user interface. Acronyms are often confusing, especially to new users. With macro programming and templates designed as a result of Chapter 3, you can design your own user interface and avert these problems.

- To enhance the overall automation of a data collection, reduction, and database management system. Using macro programs allows database management to become an integral part of a larger system that gets data, automatically formats data, performs data analyses, and then archives the results for future viewing. This method is not only safer, but minimizes technician time, thereby improving throughput.
- To provide a remedy for situations where formulae in cells or the Lotus menu commands cannot solve a problem.

This chapter will give you some macro programming basics. Although often different in format, the Lotus macro programming tools can be used to emulate virtually all BASIC commands. However, Lotus goes beyond BASIC. Lotus has tools that emulate powerful C language operations (such as pointers to variables and pointers to functions). Thus, you can think of Lotus macro programming as a hybrid between BASIC and C.

The next sections in this chapter will give you a brief overview of Lotus macro programming, followed by a comparison of the Lotus Command Language, @functions, and macro programs to BASIC and C. These discussions should get you started. This chapter also describes how to apply some more commonly used macro commands and @functions to create some simple programs.

Chapter 7 begins a series of chapters that explains how to apply these skills to the process of linking data acquisition, data analysis, and database management functions together to create a user friendly LIM system.

Entire volumes have been written on macro programming. To keep the size of this chapter small enough to allow room for the primary purpose of this book (database management), this chapter cannot be exhaustive. Therefore, the topics covered will be treated as they apply to this book. If you want to learn more about macro programming and the Lotus Command Language, you may want to review the first two books in this series (***A Complete Guide To Instrument Interfacing*** and ***Practical Spreadsheet Statistics And Curve Fitting For Scientists And Engineers***), or investigate one of the books listed in the references at the end of Chapter 1. For a quick review of the definitions for Lotus @functions and macro commands, see Appendices B and C, respectively. For a quick review of the definitions for @BASE @DB @functions, see Appendix D.

MACRO PROGRAMMING OVERVIEW

The first step in creating any macro program is to create and test a template to hold your data and data reduction summaries. Chapter 3 gave many pointers regarding template design. After your template is in place, your next step will be to write your macro program.

As stated at the beginning of this chapter, Lotus 1-2-3 and Symphony have a built-in programming language, called the Lotus Command Language. The Lotus Command Language allows you to automate virtually any procedure that you can perform manually through the keyboard. A program that uses the Lotus Command Language to automatically perform Lotus procedures is called a ***macro*** program, or simply a ***macro***.

In their simplest form, macros are time-savers. You can save time and keystrokes by creating a macro that "presses the keys" for you. That is, a simple macro contains instructions that identify which keystrokes you would use to perform a specific task. These keystrokes are stored as labels (text) within cells of your spreadsheet. When you run a macro, Lotus 1-2-3 or Symphony reads and executes the keystrokes and commands stored in the macro. For example, you can use the following macro to "type" the word "hello" into two adjacent cells (~ means "press [RETURN]" in the Lotus Command Language):

```
\R    hello~
      {RIGHT}
      hello~
```

Similarly, you can create a macro to automatically use the menu to perform certain tasks. For example, you would use one of the following macros to erase cells B9 through E14:

For Lotus 1-2-3	For Symphony
\R /REB9..E14~	\R /EB9..E14~

The Lotus Command Language has numerous programming commands that allow you to prompt users for input, perform file operations, calculate formulae, assign values to cells, transfer data from one cell to another, control the display screen, move the cell-pointer, control the path of macro execution, display custom menus, trap errors, and perform a host of other useful functions. In fact, Lotus has more than 40 specialized macro commands available (see Appendix C).

In addition to macro commands, you can create efficient programs by placing @functions within macro commands. This type of programming is called ***nesting*** and is described more fully in subsequent sections of this chapter.

Thus, the macro language is like having a built-in BASIC programming language. It is just as easy, or easier, to use and has many advantages over BASIC. These advantages will become readily apparent to you in the comparisons that will be shown in this chapter and as you proceed through the remainder of this book.

CREATING MACRO PROGRAMS

You create a macro program by typing a list of commands horizontally down a single column of a spreadsheet. These instructions are entered directly into cells just like any other data. Each instruction can be made up of one, or a combination of, the following:

- Keystrokes that you would press if you were entering data.
- Keystrokes that you would use to select items from Lotus and @BASE menus.
- Keystrokes that you would use to respond to a prompt.
- Commands that use one of the Lotus Command Language keywords.
- Commands that are a combination of Lotus Command Language keywords, @functions, and/or formulae.

To name a macro, place a name for the macro in the cell to the left of the first cell of the macro and use the /***R***ange ***N***ame ***L***abel ***R***ight command to assign a name to the macro. For example, to create the "hello" example, you would type '\R (single quote, backslash R) into a cell of the spreadsheet. To the right of the cell, you would type the program. To ***name*** the program, you would move the cell-pointer to the cell containing the '\R, issue the /***R***ange ***N***ame ***L***abel ***R***ight command and press [RETURN].

Use meaningful names for your macros. You will normally use a backslash (\) and a letter for the first macro in your program. This designation will allow you to start a chain of programs by using the [ALT] key and its corresponding letter. However, that macro will usually call upon other macros, and those macros should have descriptive names.

Choose names that relate to the processes that you are trying to achieve. Try to use names that remind you of the macro's purpose. And, above all, do not use YOUR name as a name for a macro. It can be embarrassing to come back to a program once you have become an accomplished macro writer and find your name as a macro.

Also, observe the following words of caution when you are deciding on a name: Do not use

- names that look like cell addresses (e.g., N6, AB23).
- one of the Lotus special keynames, keywords or macro names (e.g., PUT, QUIT).
- spaces or symbols within a name.
- more than 15 characters.

Sometimes you can get away with names that break the rules, but most of the time you cannot. Remember, unconventional names can lead to dangerous programming situations.

You can have as many macros in your spreadsheet as you like. You can also place them anywhere in a spreadsheet.

Although the length of a cell entry is limited to 240 characters, the size of your macro is virtually unlimited because the macro can include as many cells in a column as it takes to accomplish the task that you need it to perform. The only limitation is the amount of available computer memory.

MACRO EXECUTION

To start a macro running, specify it by name. For example, in the simple macros shown above, the macros were called by the name \R. To start a macro with this name, you would hold down the [ALT] key, press R, and then release both keys. (You would ***not*** type the backslash (\)).

If you assign a descriptive macro name in Symphony, you can invoke the macro by pressing the User key (F7), typing in the name, and pressing [RETURN].

Lotus reads program lines starting with the cell to the right of the cell containing the name of the macro and proceeds downward. (That is, Lotus starts with the cell that was specified by the /***R***ange ***N***ame ***L***abel ***R***ight command, which is the cell just to the right of the macro's name.) Lotus performs instructions in sequence. As each instruction is completed, Lotus automatically proceeds to the next cell down.

A macro terminates when it encounters an instruction to stop (the {QUIT} command), a cell entry that contains an instruction that is not valid, or an empty cell. Most macros in this book terminate with an empty cell.

PROGRAM STRUCTURE

Nearly every program that you write will require the same basic components. The first part of a program is a macro (or set of macros) that is executed at the start-up of the program. Macros of this nature ***initialize*** the spreadsheet, other programs, and your instruments (if applicable). The purpose of this macro is to combine all of these components into a system that will carry out the tasks specified by the remainder of the program.

Initialization macro(s) are important because they set all of the starting values within the system to the prescribed conditions needed by the system to operate. For example, before extended relational database operations can be performed, database files and index files need to be opened and joined. An initialization macro that has been created to auto-execute when a spreadsheet is retrieved can perform these functions to ensure that the system environment is appropriately activated prior to use. Initialization macros can also automatically execute menus of your own design, other programs, etc.

The next part of a program is the ***main*** program. This component program is the central control for the remainder of the program and will be returned to whenever the tasks in other modules have been completed. View the main program as the central hub from which subroutines are controlled. The main program in Figure 6-1 is called \R.

A subroutine is just another macro program that is apart from the main program and can be called upon by the main program. A subroutine usually performs one very specific task. When a subroutine is called, control is transferred from the main program to the subroutine, the subroutine carries out the task that it was programmed to perform, and then the subroutine returns control back to the main program. For example, some

```
------P----------Q--------R---------S---------T--------U---------V---------
1
2   \R        {GETNUMBER "HOW MANY STANDARDS?",NUM_ROWS}
3             {GETNUMBER "HOW MANY REPLICATES/STANDARD?",NUM_COLS}
4             {FOR ROW,0,NUM_ROWS-1,1,DO_ROW}
5             {CHK_MIN}
6             {BRANCH CHK_MAX}
7   FINISH    {CALC}
8
9   DO_ROW    {LET ROW_SUM,0}
10            {FOR COL,0,NUM_COLS-1,1,DO_COL}
11            {PUT STDF,0,ROW,ROW_SUM/NUM_COLS}
12
13  DO_COL    {LET ROW_SUM,ROW_SUM+@INDEX(STDSECT,COL,ROW)}
14
15  CHK_MIN   {IF @INDEX(STDF,0,0)>110}{WARN}
16
17  CHK_MAX   {IF @INDEX(STDF,0,5)<440}{BRANCH WARN}
18            {BRANCH FINISH}
19
20  WARN      {BEEP}{BEEP}
21
22
23
24
25  NUM_ROWS         6
26  NUM_COLS         4
27  ROW              6
28  COL              4
29  ROW_SUM       1878
30
```

Figure 6-1

subroutines display custom menus, print, store files, perform statistical calculations, etc. Example subroutines in Figure 6-1 are DO_ROW, DO_COL, CHK_MIN, CHK_MAX, AND WARN.

The last section of a program is known as the ***scratch-pad*** section. The scratch-pad is a set of cells that temporarily stores data. Cells Q25..Q29 constitute the scratch-pad in Figure 6-1.

View the scratch-pad section as a temporary work area ... the electronic equivalent of a note-pad. Data are placed into the scratch-pad from a variety of sources. For example, a {FOR} command will keep track of how many times its loop has been executed in one of the cells of the scratch-pad. (Programs that are executed in a repetitive manner are referred to as ***loops***.) In this context, the scratch-pad cells hold numbers that are used for counting in programs that repeat a prescribed number of times. Data that have been input by a user are also stored in the scratch-pad.

For you to fully understand the function of a scratch-pad, you must first know about the important concept of a ***buffer***. A buffer is merely a temporary storage area for data and is an excellent example of one of the primary functions of a scratch-pad. A buffer

```
  10 DIM A(10,10),B(10)
  20 DATA 93.9,92.9,94.3,94.3
  30 DATA 130.4,134.6,133.4,135.9
  40 DATA 174.5,171.8,173.1,174.1
  50 DATA 202.9,204.5,204.2,205.2
  60 DATA 235.0,242.2,242.7,239.9
  70 DATA 473.9,462.0,469.3,473.2
  80 INPUT "HOW MANY STANDARDS";NUMROWS
  90 INPUT "HOW MANY REPLICATES/STANDARD";NUMCOLS
 100 FOR ROW=0 TO NUMROWS-1
 110       ROWSUM=0
 120       FOR COL=0 TO NUMCOLS-1
 130             READ A(ROW,COL)
 140             ROWSUM=ROWSUM+A(ROW,COL)
 150       NEXT COL
 160       B(ROW)=ROWSUM/NUMCOLS
 170       PRINT B(ROW)
 180 NEXT ROW
 190 IF B(0)>110 THEN GOSUB 1000
 200 IF B(5)<440 GOTO 1000
 210 END
 220 BEEP:BEEP
 230 PRINT "ERROR!!!!"
1000 RETURN
```

Figure 6-2

is usually empty at the beginning of a program. When data need to be temporarily stored, they are stored in the buffer. When more data need to be stored, new data overwrite the current contents of the buffer cell.

As you can deduce from the description, the scratch-pad is usually in a state of constant flux. Information that is placed into cells is overwritten time after time. The scratch-pad is one of the most active portions of the entire spreadsheet and can be informative to watch.

COMPARISON OF LOTUS TO BASIC

With the above groundwork in mind, let us examine macro programming more specifically. Figures 6-1 and 6-2 show two simple programs that accomplish the same tasks. The program in Figure 6-2 is written in BASIC and the program in Figure 6-1 is an equivalent program written using Lotus tools. Both perform the relatively simple task of finding the average fluorescence value for each of the standards in Figure 6-3. Admittedly, the programming in these examples is not very well-designed or efficient, nor are the programs particularly useful. However, the examples fulfill their two primary purposes:

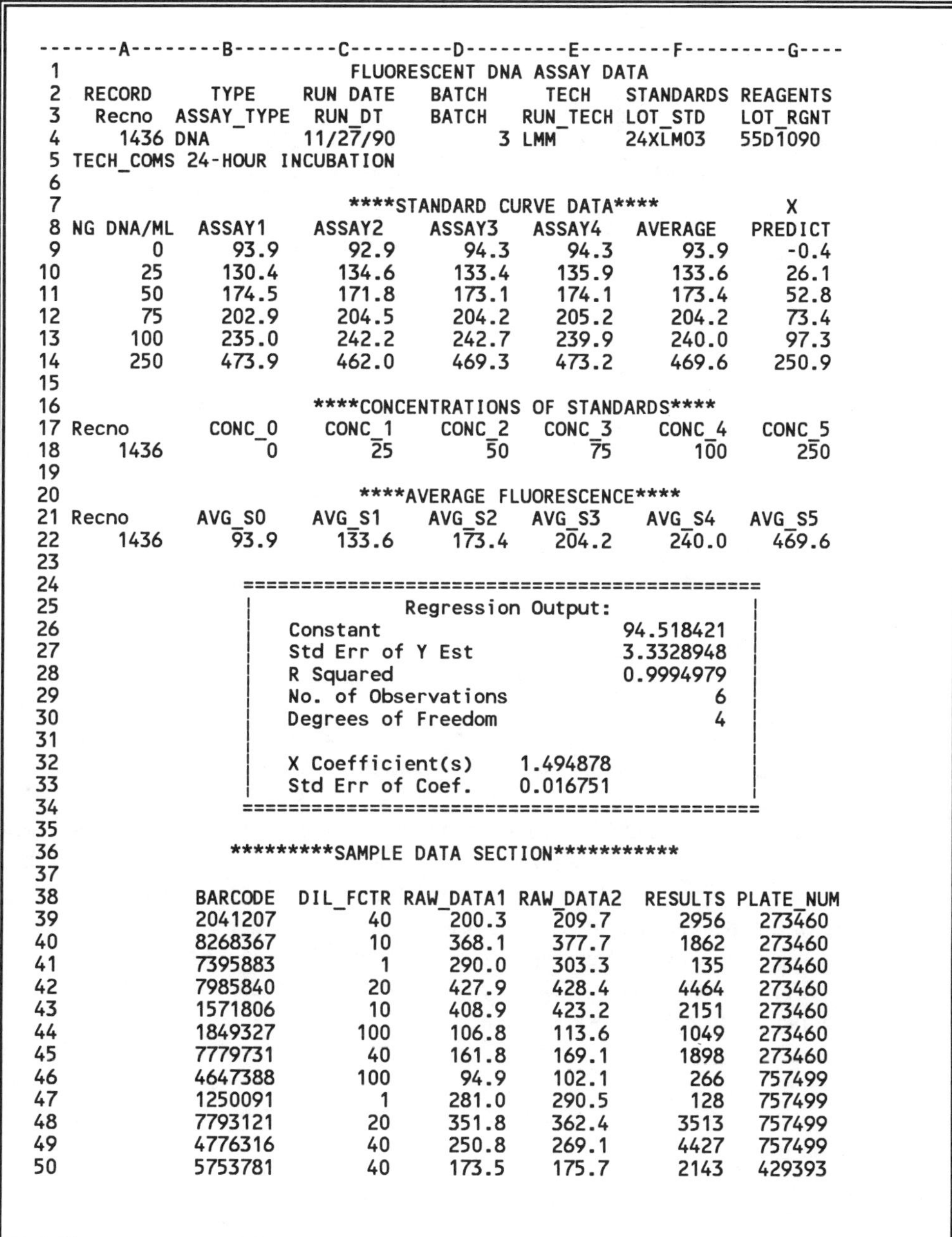

```
-------A--------B---------C---------D---------E--------F---------G----
1                         FLUORESCENT DNA ASSAY DATA
2  RECORD     TYPE    RUN DATE   BATCH      TECH   STANDARDS REAGENTS
3   Recno  ASSAY_TYPE  RUN_DT    BATCH    RUN_TECH LOT_STD    LOT_RGNT
4     1436 DNA        11/27/90          3 LMM      24XLM03    55D1090
5 TECH_COMS 24-HOUR INCUBATION
6
7                         ****STANDARD CURVE DATA****             X
8 NG DNA/ML   ASSAY1    ASSAY2    ASSAY3   ASSAY4   AVERAGE    PREDICT
9          0     93.9      92.9      94.3     94.3      93.9      -0.4
10        25    130.4     134.6     133.4    135.9     133.6      26.1
11        50    174.5     171.8     173.1    174.1     173.4      52.8
12        75    202.9     204.5     204.2    205.2     204.2      73.4
13       100    235.0     242.2     242.7    239.9     240.0      97.3
14       250    473.9     462.0     469.3    473.2     469.6     250.9
15
16                     ****CONCENTRATIONS OF STANDARDS****
17 Recno        CONC_0    CONC_1    CONC_2   CONC_3    CONC_4    CONC_5
18      1436         0        25        50       75       100       250
19
20                       ****AVERAGE FLUORESCENCE****
21 Recno       AVG_S0    AVG_S1    AVG_S2   AVG_S3    AVG_S4    AVG_S5
22      1436     93.9     133.6     173.4    204.2     240.0     469.6
23
24                ============================================
25                |           Regression Output:             |
26                |   Constant                    94.518421  |
27                |   Std Err of Y Est            3.3328948  |
28                |   R Squared                   0.9994979  |
29                |   No. of Observations                 6  |
30                |   Degrees of Freedom                  4  |
31                |                                          |
32                |   X Coefficient(s)    1.494878           |
33                |   Std Err of Coef.    0.016751           |
34                ============================================
35
36               *********SAMPLE DATA SECTION***********
37
38            BARCODE  DIL_FCTR RAW_DATA1 RAW_DATA2  RESULTS PLATE_NUM
39            2041207        40     200.3     209.7     2956    273460
40            8268367        10     368.1     377.7     1862    273460
41            7395883         1     290.0     303.3      135    273460
42            7985840        20     427.9     428.4     4464    273460
43            1571806        10     408.9     423.2     2151    273460
44            1849327       100     106.8     113.6     1049    273460
45            7779731        40     161.8     169.1     1898    273460
46            4647388       100      94.9     102.1      266    757499
47            1250091         1     281.0     290.5      128    757499
48            7793121        20     351.8     362.4     3513    757499
49            4776316        40     250.8     269.1     4427    757499
50            5753781        40     173.5     175.7     2143    429393
```

Figure 6-3

BASIC	LOTUS
INPUT "message";NUMROWS	{GETNUMBER "message",NUM_ROWS}
FOR ROW=0 TO NUMROWS-1	{FOR ROW,0,NUM_ROWS-1,1,DO_ROW}
ROW_SUM=0	{LET ROW_SUM,0}
ROW_SUM=ROW_SUM+A(COL,ROW)	{LET ROW_SUM,ROW_SUM+ @INDEX(STDSECT,COL,ROW)}
IF B(0)>110 THEN GOSUB 1000	{IF @INDEX(STDF,0,0)>110}{WARN}
GOSUB 500	{CHK_MIN}
GOTO 600	{BRANCH CHK_MAX}
B(ROW)=ROW_SUM/NUMCOLS	{PUT STDF,0,ROW,ROW_SUM/NUM_COLS}
BEEP	{BEEP}
No equivalent	{WAIT @NOW+@TIME(0,0,5)}

Figure 6-4

- To quickly and briefly illustrate some of the more important similarities between Lotus and BASIC.
- To give you examples of how to use macro commands and @functions to move data around a spreadsheet.

Before we get into the details of the programs, take a moment to look qualitatively at the two programs and compare them. Note the differences in their format. For example, in BASIC, line numbers are needed to keep track of program flow (e.g., 10, 20, etc.). The program is contiguously aligned. Neither requirement applies to Lotus. Line numbers are not used in Lotus. As stated above, Lotus reads down a column of cells and senses the end of a macro program or subroutine by an empty cell. Also note that each subroutine has a name in Lotus, but a line number in BASIC.

Also note the differences in the method of navigating around the program. In BASIC, the GOTO and GOSUB commands are used; in Lotus, {BRANCH subroutine} and {subroutine} are the equivalent commands.

Finally, in BASIC, all of the variables are either read in from DATA statements or are hidden from routine view. If a matrix is required, a DIMension statement specifies the size of the matrix. When the elements of the matrix are referred to, they are specified using the following formats:

A(col,row)
B(row)

In Lotus, variables are stored in the cells of a spreadsheet. Temporary storage is often located in scratch-pad buffers that are always available for review. Placing data into individually specified cells is executed via {LET} commands. If a matrix is needed, no DIMension statements are required because the matrix is comprised of cells in the spreadsheet. ***To get data into a matrix, use the {PUT} command. To retrieve data from a matrix, use the @INDEX function.*** Both the {PUT} command and the @INDEX function take a named range, a column offset, and a row offset as specifications for the cell in the matrix being referred to.

Next, take a moment to look at the syntax of some common commands. Figure 6-4 is a side-by-side list of the differences in Figures 6-1 and 6-2. As you can see, although these two languages have slight differences, they are actually very similar. If you are a beginner, this similarity will make it easy for you to "translate" programs in text books into useful programs that manage data in your Lotus templates.

The Lotus program in Figure 6-1 contains the tools and techniques most often used in general programming. For this reason, the remainder of this chapter is dedicated to giving details on the tools and techniques used in Figure 6-1.

OBTAINING INFORMATION FROM A USER

The main program (\R) of Figure 6-1 illustrates one of the methods that is available for prompting the user for input. In this example, a pair of prompts asks the user to enter the number of standards and the number of replicates/standard.

The command that you use to obtain numbers from a user is the {GETNUMBER} command. For text, use the {GETLABEL} command. The syntax of these two commands is

```
{GETNUMBER prompt,location}
{GETLABEL prompt,location}
```

When either of these two commands is executed in a macro, a prompt is displayed on the second line of the control panel and execution of the macro is paused until input is made. When the user presses [RETURN] on the keyboard, the input is placed into the cell specified by "location". In the example program, the number of standards goes into the cell named "NUM_ROWS" and the number of replicates goes into the cell named "NUM_COLS".

User input obtained from a {GETNUMBER} or {GETLABEL} command can be used later by your program. For example, in this program the "NUM_ROWS" number will set the limit for the main loop counter and the "NUM_COLS" number will set the limit for a second loop counter. (A description of how this information is incorporated is given below.)

Another use of the {GETLABEL} command is to pause a program at specific points during execution. That is, by strategically placing a {CALC} command and a {GETLABEL "PAUSE",TXT} command in your program, you can temporarily stop the program to view the current status of a template, scratch-pad, etc. When you press [RETURN], the program will continue until another {GETLABEL} command is encountered. This use makes {GETLABEL} a very valuable learning aid and debugging tool.

An alternative programming tool that you can use to get input from a user is the {?} command. The {?} command halts macro execution, thus allowing the user to type any combination of keystrokes, function keys, or arrow keys (to move the cell-pointer). The macro will continue after the user presses [RETURN]. This pause allows the user to input text or numbers directly into cells anywhere in the spreadsheet.

Pressing [RETURN] has no effect other than to end the {?} command. That is, pressing [RETURN] with the {?} command activated will NOT permanently enter data into a cell. To permanently enter user input, you must include a tilde (~) in the cell, after the {?}. A tilde means "Press [RETURN]." Without a tilde after the {?} command, the data that were input may be lost.

DATA REDUCTION USING LOOPS

An example of nested {FOR} loops is shown in the Figure 6-1 program. Nested {FOR} loops are perhaps the most commonly used methods for data acquisition, data analyis, and database management. You will probably use them often for these chores, so we will take a closer look at how they work.

The outer {FOR} loop in the main program simply executes the inner {FOR} loop NUM_ROWS-1 times. One is subtracted because loop execution starts at zero, not one. The zero starting point will become more obvious later when we discuss the offsets for @INDEX functions and {PUT} commands.

The format of the outer loop is:

```
{FOR ROW,0,NUM_ROWS-1,1,DO_ROW}
```

where,

DO_ROW is the name of the macro that you want the spreadsheet to execute repeatedly, and the arguments ROW, 0, NUM_ROWS, and 1 represent the scratch-pad cell that will contain the current loop number, starting value, scratch-pad cell containing the stop value, and the step size, respectively.

The "nested" (inner) {FOR} loop is in the DO_ROW subroutine. After ROW_SUM is reset to zero, the {FOR} loop executes the DO_COL subroutine NUM_COLS-1 times. DO_COL performs a summation on each of the fluorescence values in a standard's row.

By constructing this combination of commands for the standard section in Figure 6-3, you can get an average of replicate readings for each standard in the standards section. The results are placed into sequential rows of the STDF range of the spreadsheet template (based on the ROW counter of the first {FOR} loop).

GETTING DATA FROM A TEMPLATE

The @INDEX function in the {LET} command of DO_COL retrieves data from a template matrix.

@INDEX is one of the most useful @functions in a class of @functions that Lotus calls ***lookup @functions***. The @functions in this class take a table and look up one of the pieces of datum in the table based on its position. Thus, these @functions require a definition of the table AND the offsets that pinpoint the piece of datum that you want.

The @INDEX function is a very handy programming tool. The tables that the @INDEX function uses are named ranges in a spreadsheet. The syntax of @INDEX is

```
@INDEX(range,column_offset,row_offset)
```

This @function pinpoints a cell located within a range based on specified column and row offsets and returns the value of the cell. Using @INDEX, you can access any cell within a two-dimensional array (matrix) and get its contents. In this example, STDSECT has been defined as cells B9..E14. The column and row offsets in an @INDEX function begin at (zero,zero). This starting position is one of the reasons for starting the {FOR} loop counters at zero and not one. The @INDEX in Figure 6-1 has the following format:

```
@INDEX(STDSECT,COL,ROW)
```

Thus, as the COL and ROW counters in the two {FOR} loops increment, they specify the column and row offsets for @INDEX. Using COL and ROW as the column and row offsets allows you to increment the offsets and thereby change the data being utilized in the calculation. For example, when COL and ROW are both zero, @INDEX takes on the value of cell B9 (column-offset=zero, row-offset=zero). In Figure 6-3, that value is 93.9. When COL increments to 1, @INDEX takes on the value of cell C9 (column-offset=one, row-offset=zero). In Figure 6-3, the @INDEX value is 92.9. This process of advancing through the matrix using loop counters continues for the duration of both loops.

MOVING DATA INTO A TEMPLATE

After a full row's summation has been completed, the program uses a {PUT} command to transfer the data into the appropriate cell of the template. The format of the {PUT} command is

```
{PUT range,col_offset,row_offset,value}
```

where,

range defines the block of cells in which the {PUT} could place the result; col_offset specifies the column (left/right) offset within the range; row_offset specifies the row (up/down) offset within the range; and value is the result that is to be placed into the cell specified by the column and row offsets. Value can be a cell address or the name of a cell containing data, or it can be a formula, @function, etc. As in the @INDEX case, the first row and column of the range have the offset (zero,zero).

Thus, the {PUT} macro command pinpoints a cell located within a range based on specified column and row offsets and places a value into the cell. Using {PUT}, you can change the value of any cell within a two-dimensional array (matrix). In the example program of Figure 6-1, STDF has been defined as cells F9..F14. Because the column offset in a {PUT} command begins at zero, the column offset has been permanently set to zero. That is, it is a one-dimensional array (i.e., a column matrix).

The row offset in this example program comes from the outer loop counter. The row offset for a {PUT} command begins at zero. This starting position is the second reason for starting the {FOR} loop counters at zero. The {PUT} command in Figure 6-1 has the following format:

```
{PUT STDF,0,ROW,ROW_SUM/NUM_COLS}
```

Thus, as the ROW counter in the outer {FOR} loop increments, it specifies the row offset for the {PUT} command. Using COL and ROW as the column and row offsets thereby allows you to increment offsets and change the cell being modified. For example, when COL and ROW are both zero, {PUT} places the value of ROW_SUM/NUM_COLS into cell F9 (column-offset=zero, row-offset=zero); when ROW increments to 1, {PUT} places the value of ROW_SUM/NUM_COLS into cell F10 (column-offset=zero, row-offset=one); and so on.

The {PUT} command has a built-in feature that can add a measure of safety to your program. Notice that I used a single column (F) as the range of acceptable cells in the {PUT} command. This specification provides a safety measure because it does not allow the program to place data in sections of the template or program where it would overwrite formulae and/or macro commands. This feature is an important one that you will want to exploit in your programs.

{PUT} commands provide a very desirable function and using them in place of template @functions should be considered often. As stressed at the beginning of this chapter, @functions and formulae slow the recalculation speed of spreadsheets. If @functions are nested within a macro {PUT} command, the calculation would only be performed at the time that the {PUT} command was executed. This method can therefore improve the recalculation speed of your spreadsheet.

CALLING AND BRANCHING TO MACRO PROGRAMS

From a macro program, you can execute a second macro program in one of two ways. The first way is to call the second macro as a subroutine. If the second macro is called as a subroutine, control of the program is temporarily shifted to the subroutine, the subroutine carries out the tasks it was programmed to perform, and then control is returned to the original, calling program.

The main program in Figure 6-1 calls CHK_MIN as a subroutine. Notice the method that is used to call the subroutine. ***Whenever you call a subroutine from either a macro or menu, place the name of the subroutine within { } braces.*** For example, to call CHK_MIN, type the following text into your program:

```
{CHK_MIN}
```

If your macro has a backslash-letter name (e.g., \R), then type the following text into your program to call the macro subroutine:

```
{\R}
```

In the example, after the CHK_MIN program has completed its execution, program control returns to the main program.

The second way to execute a macro is to branch to it using the {BRANCH} command. The {BRANCH} command permanently transfers control of the program to the cell named in the {BRANCH} command. That is, when a macro is executed with the {BRANCH} command, the program jumps to the new macro. Execution terminates in the original macro and total control transfers to the new macro. Then, the new macro executes. When the last instruction in the new macro is encountered, program execution will terminate (unless another branching command is encountered).

Branching is illustrated in the main program of Figure 6-1. The example is

```
{BRANCH CHK_MAX}
```

Note that the CHK_MAX macro has a {BRANCH FINISH} command. This command is required to return to the main program. Without it, the program would terminate when CHK_MAX had completed its tasks. Program termination is illustrated by the branching to WARN when the {IF} test in CHK_MAX is true. Because WARN is not called as a subroutine and does not have a {BRANCH} command, the program would terminate via this pathway.

There is a subtle difference in these two methods of executing macros. On the surface, the main difference is that a subroutine call will automatically return to the calling program, while a {BRANCH} will not. However, the ramifications of this subtle difference can be very important to the reliability of a program's execution because Lotus limits the number of subroutine levels that you can call while within a program.

If a macro calls another macro as a subroutine, Lotus is supporting two ***levels*** of subroutines. Likewise, if a second subroutine calls a third subroutine, Lotus is supporting three levels, and so on. The number of subroutine levels that either Lotus 1-2-3 or Symphony can support (without returning to the first macro level) is 31.

If you exceed the subroutine limit by calling a macro from itself and/or by trying to nest too many subroutines, the macro aborts and an error message appears.

Whenever possible, I prefer {BRANCH}ing as opposed to subroutine calls because {BRANCH}ing allows more levels of macro nesting, thereby allowing programs to be longer and more efficient.

{IF} TESTS

The {IF} tests in the CHK_MIN and CHK_MAX macros check to ensure that the fluorescence units are within acceptable limits. This programming technique may be required by an assay to ensure that the blank is not too high, that there is sufficient sensitivity, etc. However, the real intention of the example is to illustrate the {IF} command.

An {IF} command allows a macro program to perform a specific task based on the result of a TRUE/FALSE test. If the result of the expression within the {IF} command does not have a numeric value of zero, Lotus considers the expression to be TRUE and the macro executes the instructions immediately after the {IF} command (***in the same cell***). If the result of the expression within the {IF} command has the numeric value zero, Lotus considers the expression to be FALSE and the macro continues ***to the cell below*** the one containing the {IF} command.

In this example, the two {IF} tests call the subroutine {WARN} if the argument specified in the command is TRUE. If the argument to the {IF} test is FALSE, the program just skips the call to the WARN subroutine and proceeds down to the next command line. You can place any Lotus macro command, subroutine call, menu command, branch, etc., after an {IF} command.

FORCING SPREADSHEET RECALCULATION

The {CALC} command at the end of the program forces a recalculation of all formulae in the spreadsheet and updates, or re-displays, the current data. When a macro program executes, it does not necessarily recalculate the formulae in a template. ***Placing a {CALC} command at the end of a macro is very important to avoid user confusion.***

More importantly, if a user forgets to recalculate a spreadsheet before printing it, the results may not be correct because they may not reflect new values that have been added to the template by the macro program. To prevent the possibility of catastrophe, get into the habit of placing a {CALC} command at the end of all of your macro programs and, more importantly, before any automated printing is performed by your program.

THE SCRATCH-PAD

Earlier in this chapter, we looked briefly at the concept of a scratch-pad. A scratch-pad is a group of cells that temporarily stores data and holds values for loop counters. Your program relies very heavily on the data in your scratch-pad. This relationship makes the scratch-pad an important part of your program and makes it an area where you can greatly enhance the efficiency of your program.

For example, in Figure 6-1 COL and ROW serve three purposes. Their primary function is to serve as loop counters. When the {FOR} loops in this example begin execution, they set the value of their loop counters to zero. Each time a loop is repeated, the {FOR} command increments the counter by one. Therefore, a {FOR} loop needs a cell in the spreadsheet to hold the current loop value. COL and ROW fulfill this requirement.

The second function of the loop counters is that they serve as the column and row offsets for an @INDEX function. As COL and ROW increment with the loops, the appropriate cell in the range is designated and the @INDEX takes on the value of the cell.

The ROW loop counter has a third function. It specifies the row offset values in the {PUT} command. Thus, as the ROW counter in the outer {FOR} loop increments, it automatically increments the row coordinates within the STDF range and the average data are placed in the specified cells. This system of {FOR} loops, @INDEX functions, and {PUT} commands is a very handy way to position data and one that you will be using often.

The COL and ROW cells make neat programming packages, tying everything together. This kind of package programming is very efficient and decreases the chance of error.

Another important feature of the scratch-pad is the cell called ROW_SUM. We discussed the concept of a buffer earlier in this chapter, but it seems appropriate to repeat some of the discussion as it applies to the scratch-pad. If you recall, a buffer is a temporary storage area for data. It is usually empty at the beginning of a program. When data come in from a program, they are stored in the buffer. When more data come in, they overwrite what was previously in the buffer.

In the example program, ROW_SUM is initialized to zero each time the outer {FOR} loop increments and calls DO_ROW. Thereafter, each time DO_COL is called by the inner {FOR} loop, the existing data in ROW_SUM are queried to find the existing summation, the new value is added to it, and the old value is overwritten with the new value.

User input frequently requires the use of a buffer. For example, the {GET-NUMBER} and {GETLABEL} commands require a single cell specification (e.g., NUM_ROWS and NUM_COLS). This requirement means that a buffer cell is needed to receive the data. Therefore, if you use one of these commands to input data for template cells, you need to set up a buffer to receive the data. Once received, you can transfer the data to any cell of the spreadsheet with a {PUT} command. When subsequent user input is made, the old data are overwritten with new, processed, and so on.

Another common use of buffers is to obtain data from instruments. In my previous book, ***Laboratory Lotus®: A Complete Guide To Instrument Interfacing***, buffers were required whenever data were acquired from an instrument. In this context, the buffer served as a staging area for {READLN} commands. {READLN} needs a specific location to place the data that are read. {READLN} cannot pinpoint data with column and row offsets the way that {PUT} can. Once data have been placed in a buffer, they are parsed with @functions and then moved to another cell in the spreadsheet with a {PUT} command.

GETTING ACQUAINTED WITH CREATING MACROS

Even though you will not be using the example application in a "real life" situation, type the program in Figure 6-1. Use the template spreadsheet that you created in Chapter 3. ***Make certain that the macro is located to the right of the last column of cells in the template.*** Also, make sure that you leave at least one blank row between each macro; Lotus 1-2-3 and Symphony look for these blank rows to determine the end of a macro.

It is easy to prepare a section of the spreadsheet for your scratch-pad: Just type the names of the variables into the same column as the macro names.

The next task in creating any program is to "activate" the program and scratch-pad. To accomplish this task, assign label names to the macros and scratch-pad cells by issuing the /***R***ange ***N***ame ***L***abel ***R***ight command. When the spreadsheet prompts you for the range of labels, press [ESC]ape, move the cell-pointer to the cell containing the first macro name, type a period (.), use the [DOWN] arrow key to highlight all of the macro names AND the scratch-pad names, and press [RETURN]. This procedure defines the cells to the right of the column as having the names specified in the name column. You should now be able to appreciate the convenience of aligning range names into a single column.

Finally, give cells B9..E14 the name STDSECT. Do so by issuing the /***R***ange ***N***ame ***C***reate command. When the spreadsheet prompts you for the name, type STDSECT, and press [RETURN]. When prompted for the range of cells, type B9..E14 and press [RETURN]. Repeat this process to give cells F9..F14 the name STDF.

Now, save the spreadsheet. To execute the program, press [ALT]-R.

TEST IT!

It is better to test macro programs as you create each section rather than to wait until the entire program has been completed. If you wait until the end of the programming process, the many subroutines and program lines will make it more difficult to isolate errors. However, if you test as you build, you can verify that each section works correctly before you add another level of complexity.

Test your macro program with actual experimental data. This real-life testing often uncovers problems that cannot be found by just testing the system with contrived data.

After you have completed the entire program, perform a final, exhaustive test of the complete system. A template and macro program that have not been thoroughly tested with actual experimental data should not be used to report results. Important decisions may be based on a report that looks formal, but contains erroneous results. For this reason, you should compare the results of the macro program calculations in your template with the results that you obtain from a calculator, etc.

BACK UP SEVERAL COPIES

Once you have added all of your enhancements and completed final testing, your spreadsheet will represent a considerable time investment. You must protect this investment.

Save the spreadsheet (without data) on several, clearly labeled floppy disks. Also, print out the program for hard copy documentation. Save all of the items in several different, safe places.

MAKING YOUR JOB A LITTLE EASIER

The remainder of this chapter covers topics that will help you create more professional programs. This section will cover the topic of using named ranges to achieve program flexibility.

Whenever you prepare a program, make the programming as flexible as possible to allow for future modification. That is, if you need to modify a program in the future, you will want to be able to implement the update with a minimum of changes. As a general rule, the fewer changes required to implement an update, the less chance of forgetting to edit something that will affect the program's performance. One very good programming technique that can be of help achieving this goal can be seen in the {PUT} commands and @INDEX functions that appear in the DO_ROW and DO_COL macros of Figure 6-1.

If you recall, earlier in this chapter named ranges were placed into a {PUT} command and @INDEX functions. Rather than using STDSECT and STDF in the program, you could have specified B9..E14 and F9..F14, respectively.

However, if you change a template (by moving cells, inserting or deleting rows or columns, etc.) and if the {PUT} and {LET} commands and/or nested @functions of your program contain cell addresses, all the addresses would need to be edited. In contrast, if you change the cell address specifications of named ranges and if named ranges have been specified in macro commands or @functions nested into macro commands, then you would not have to change a thing. For example, {PUT B9..E14, ... } would need to be edited if you moved the range over by just one column, but {PUT DATASECT, ... } would not. If you have a large program with many macro commands, editing is not only a massive amount of work, but a prime target for bugs in your program.

There is another reason why you should get into the habit of using named ranges instead of cell addresses for your {PUT} and {LET} commands and @functions: Using named ranges will make your program easier to read and follow. "STDSECT" is certainly more informative than "B9..E14". Names become very important in large spreadsheets and macro programs.

A FEW COMMENTS ON COMMENTING

Add comments to your programs for one VERY important reason: Comments make it easier for someone (including yourself) to read and understand your program. Just try to return to a program after several weeks and attempt to remember what you were trying

```
------P----------Q--------R---------S---------T--------U---------V---------
1  /*PROGRAM TO FIND AVERAGE FLUORESCENCE UNITS FOR STANDARD CURVE*/
2   \R       {GETNUMBER "HOW MANY STANDARDS?",NUM_ROWS} /*REV: 11/19/1990*/
3            {GETNUMBER "HOW MANY REPLICATES/STANDARD?",NUM_COLS} /*LMM*/
4            {FOR ROW,0,NUM_ROWS-1,1,DO_ROW}    /*OUTER LOOP*/
5            {CHK_MIN}          /*SEE IF BLANK VALUE LOW ENOUGH*/
6            {BRANCH CHK_MAX}  /*SEE IF ENOUGH SENSITIVITY*/
7   FINISH   {CALC}             /*FORCE RECALCULATION OF SPREADSHEET*/
8
9   DO_ROW   {LET ROW_SUM,0}   /*INITIALIZE BUFFER TO ZERO*/
10           {FOR COL,0,NUM_COLS-1,1,DO_COL}   /*FIND THE ROW'S SUM*/
11           {PUT STDF,0,ROW,ROW_SUM/NUM_COLS} /*PUT AVERAGE INTO TEMPLATE*/
12                              /*ADD CURRENT CELL TO SUMMATION*/
13  DO_COL   {LET ROW_SUM,ROW_SUM+@INDEX(STDSECT,COL,ROW)}
14
15  CHK_MIN  {IF @INDEX(STDF,0,0)>110}{WARN}
16
17  CHK_MAX  {IF @INDEX(STDF,0,5)<440}{BRANCH WARN}
18           {BRANCH FINISH}
19
20  WARN     {BEEP}{BEEP}       /*SOUND SPEAKER*/
21
22
23          /*SCRATCH PAD*/
24
25  NUM_ROWS        6           /*USER DEFINED NUMBER OF STANDARDS*/
26  NUM_COLS        4           /*USER DEFINED REPLICATES/STANDARD*/
27  ROW             6           /*ROW COUNTER FOR OUTER LOOP*/
28  COL             4           /*COLUMN COUNTER FOR INNER LOOP*/
29  ROW_SUM      1878           /*SUMMATION ACCUMULATOR*/
30
```

Figure 6-5

to accomplish in a certain section of the program! Or try to remember a pitfall that you thought about while you were programming. If you document a program section at the time that you write it, the program will be at its freshest in your mind and the comments will make the best sense. In other words, comments will be more "readable" and it will be easier to correct or modify a program, if necessary.

Commenting your program also helps clarify your own concept of what the program does and often exposes design flaws. ***When commenting, try to be brief. Remember, memory utilization is at a premium in LIM system applications. @BASE add-in application programs, macro programs, indexes, and spreadsheet data tend to consume large amounts of memory. This consumption means that you must use the remaining memory efficiently. Character for character, comments use the same amount of memory as program lines. Therefore, you must be concise in your comments to try to conserve memory.*** Because the programs in this book are oriented toward teaching you programming concepts, the programs tend to be over-commented. In most applications, commenting would be less, and only comments of strategic importance would be used.

@LOG(number)	Log of number, base 10
@LN(number)	Log of number, base e
@EXP(number)	The number e raised to the number power
@SIN(number)	Sine of number
@COS(number)	Cosine of number
@TAN(number)	Tangent of number
@ASIN(number)	Arc sine of number
@ACOS(number)	Arc cosine of number
@ATAN(number)	Two-quadrant arc tangent of number
@SQRT(number)	Positive square root of number

Figure 6-6

What is the best way to comment your program? No best way exists because there is so much flexibility in what you can do. Implement a system that seems the most comfortable to you and stay with the system. Recall that when Lotus reads a macro, it reads straight down a column. If a comment is placed just one column over, it will be ignored. The `'/*comment*/` format that you see in this book is the standard way to comment in C. Figure 6-5 shows an example of how to comment using this convention.

You can use any commenting format that you like, as long as the comments are not in a cell within a program. Because macro execution begins at the cell that corresponds to the macro's name and proceeds downward, any text above the cell will be ignored. Just make sure that you leave at least one row between the comment and the macro above it. Otherwise, Lotus will think that the comment is part of the previous macro.

Another commenting practice to follow is to write a ***brief*** synopsis of the function of a program at the beginning of the program. A good synopsis contains the objectives of the program, type of testing performed, textbook references, name of the curve fitting algorithm, revision dates, programmer's initials, etc.

HANDLING SCIENTIFIC FORMULAE

Scientific data often require the use of logarithms, natural logs, exponentials, sines, cosines, etc. You can handle these cases by adding the appropriate Lotus mathematical @function(s) into the {PUT} command that places the data into cells. Figure 6-6 is a list of common mathematical @functions.

For example, to transform the value of a BUFFER cell using the @LOG function and place the transformed value into a cell in the range called ABSDATA of a template, use the following {PUT} command:

```
{PUT ABSDATA,COL,ROW,@LOG(BUFFER)}
```

NOT FOR C PROGRAMMERS ONLY

As noted at the beginning of this chapter, Lotus has tools that emulate pointers. The @@(CELL_REF) function is the redirection operator for cell contents. The redirection @function returns the contents of the cell referenced by CELL_REF. That is, the range name or cell location that appears in the CELL_REF cell is used as the address of a cell. The value returned by the @@ function will be the contents of the cell at this address. For example, if the following formula were typed into cell Q30 of Figure 6-1:

```
@@(CELL_REF)
```

and if the contents of the cell called CELL_REF were the label "ROW_SUM", the value of the ROW_SUM cell would appear in Q30.

The {DISPATCH macro_spec} macro command is the Lotus equivalent of a pointer to a function (subroutine) in C. When encountered in a macro program, the {DISPATCH} command causes a {BRANCH} to the macro that begins at the macro's name specified in the MACRO_SPEC cell, and the new macro begins executing. For example, suppose that a cell named MACRO_SPEC was placed in Figure 6-1, that the contents of MACRO_SPEC were the label "WARN", and that the following command was issued in the \R macro:

```
{DISPATCH MACRO_SPEC}
```

When the {DISPATCH} command was encountered, {DISPATCH} would branch to the macro named WARN and begin execution.

SUMMARY OF WHAT HAS BEEN ACCOMPLISHED

This chapter compared and contrasted common BASIC, C, and Lotus commands and presented an overview of program design techniques that accomplish some of the more common tasks required by nearly every data reduction macro program. To this end, a number of fundamental macro programming concepts were given:

- How to determine when to use macro programs vs. active formulae in cells.
- What a macro program is and what it is comprised of.
- How to design and create a macro program.
- How to execute macro programs.
- How to temporarily and permanently transfer program control from one macro to another.
- How to implement a scratch-pad.
- What a buffer is, its importance, and how it is used.

- How to get information ***from*** a template using @INDEX.
- How to move data ***into*** a template using {PUT} and {LET} commands.
- The importance of using named ranges in macro commands.
- The importance of testing macro programs and templates.
- The importance of commenting macro programs.

To review the concepts of this chapter, create a macro program that calculates an average for matched cells of the sequential rows of RAW_DATA1 and RAW_DATA2 of Figure 6-3 and place concentration values into the corresponding cells of the RESULTS column. This program will simulate the performance of calculations prior to transfer of data to a database. The following equation can be used:

```
(DILUTION FACTOR)*(AVERAGE-INTERCEPT)/SLOPE
```

The Lotus equivalent is:

```
@INDEX(DATA,0,ROW)*(((@INDEX(DATA,1,ROW)+@INDEX(DATA,2,ROW))/2)-INTERCEPT)/SLOPE
```

To use this equation in a {PUT} command, you will need to define the ranges for DATA, INTERCEPT, SLOPE, and RESULTS as C39..E500, F26, E32, and F39..F500, respectively. The program should be named '\C.

WHAT'S NEXT?

Now that you have learned some basic programming skills, the next chapter will begin showing you how to design and build fully functional LIM system database management programs. Besides being a very useful program in its own right, the program presented in the next chapters should give you a good example of programming techniques that you can use when you create your own programs.

7

Establishing System Sequence

Creating a LIM system is like making an article of clothing. All of the lists, templates, database files, index files, join definitions, computed fields, and view files that you forged in the previous chapters are like pattern pieces of fabric. The thread that ties the pattern pieces together is a macro program. And the guide that you use to show you how all of the pattern pieces go together is a menu system of your own design. A menu system coordinates all of the functionality of your LIM system. Therefore, a menu system is key to the development of reliable and efficient macro programs. To this end, menus serve three purposes:

- Before programming, as a tool to define specifications for the flow of the system. Typically, programmers create flow diagrams for their software. However, it is much easier to visualize the flow of software activity by creating a tree that describes a hierarchy of interconnected menus than by creating flow diagrams. The end result is the same. That is, if you design a menu system based on how you plan to use the system, you have designed the flow of the system.
- During programming, as a tool that separates subroutine interactions. Because each function selected from a menu is relatively independent, a menu system allows you to create independent subroutines without having to worry about other subroutines in the system. This narrow-minded approach to programming causes less confusion and fewer program errors. The approach also leads to more efficient programming. When creating a menu tree, just place the names of the macro programs into the appropriate cells of the menu system. Later, use the macro program names and menu help line descriptions (help messages) as specifications for the names and functionalities of the macro programs to be

created. This approach cuts even large programming tasks into small, manageable ones.

- After programming, as a user friendly tool for users to control the LIM system. The subroutines that you create are like puppets on a string being controlled by the menu system. When a user wants a certain function, a choice can be made from a menu. After selection is made, the appropriate subroutine executes and provides the functionality specified. When the subroutine has completed its tasks, control is returned to the menu so that the user can make another selection.

The next sections explain how to create macro programs that automatically execute at system start-up, configure the system, and launch a menu system. The remainder of the chapter explains how to design and create menu systems. Chapter 8 shows you how to implement menu choices in the example menu system.

AUTO-EXECUTING MACROS

You can program Lotus 1-2-3 and Symphony to automatically start a macro running when the worksheet that contains the macro is retrieved. This special macro is the first one that you would normally create whenever you start programming a new application.

A macro program that automatically executes when a spreadsheet is retrieved is called an ***auto-executing macro***. Auto-executing macros are very commonly used to perform tasks that ***must*** be accomplished ***before*** the template or other macros in the spreadsheet can be used successfully. For example, auto-executing macros are frequently used to place starting values into key cells of a template.

However, the most common and important use of auto-executing macros is to coordinate the process of creating an environment that can be used by your program. For example, before @BASE can access data in databases, the database files must be opened. This task is a prime candidate for an auto-executing macro. Another prime candidate is starting up a user menu.

These processes are referred to as ***initialization*** processes and will be explained in detail in the next few sections. The example auto-executing macro programs shown in this chapter are auto-executing macros that aid in configuring a program to run under both Lotus 1-2-3 and Symphony spreadsheets.

DIFFERENCES BETWEEN LOTUS 1-2-3 AND SYMPHONY AUTO-EXECUTING MACROS

Lotus 1-2-3 and Symphony handle most macros identically. However, auto-executing macros are handled in slightly different manners by the two spreadsheets. For example, Lotus 1-2-3 requires that an auto-executing macro be given a special, unique name ('\0). Symphony allows any legitimate macro name for an auto-executing macro as long as you tell Symphony the name.

You can include both macros in the same worksheet and the two macros will not interfere with each other. That is, if you have Symphony and have specified the name AUTOEXEC as the auto-executing macro in a spreadsheet's settings, Symphony will ignore the '\0; while Lotus 1-2-3 will only recognize '\0.

If you are writing programs that are designed to run within both Lotus 1-2-3 and Symphony spreadsheets, including two different auto-executing macros in the same spreadsheet is a good way to ensure ***portability***.

As you can tell from the previous discussions in this book, various versions of @BASE and Releases of the two Lotus spreadsheets have minor differences in the way that some of the menu commands are issued; but macro commands and @functions are identical. Therefore, if you can detect which Release of Lotus is using a macro program, you can issue the appropriate menu commands. In other words, you can exploit differences in Lotus 1-2-3 and Symphony Releases to implement a system that will allow a spreadsheet to be independent of Lotus Release. This system, in turn, allows you to share your program with colleagues or upgrade from Lotus 1-2-3 to Symphony without performing an extensive rewrite of a program. This flexibility will allow a spreadsheet to be moved back and forth from one Lotus program to the other.

With this system, you may need to re-define a few graphics or print settings, re-specify the name of the auto-executing macro (if stored in Lotus 1-2-3 and used in Symphony), rename the file extension (e.g., from ".wk1" to ".wr1" when you go from Lotus 1-2-3 to Symphony), etc., but you will not have to change the program.

To utilize this portability, you must first make certain that the auto-executing macro for Lotus 1-2-3 is named '\0 and the auto-executing macro for Symphony has another, different name. Once you have done so, you can use either of two different methods to customize subsequent macro programs. These methods are shown in Figure 7-1 and are described in more detail below.

Because of the slight differences in the two Lotus programs, discussions concerning auto-executing macros are separated by section. The next section describes the auto-executing macro for Lotus 1-2-3 and the section following describes the auto-executing macro for Symphony. Again, the difference in auto-executing macros is a special case. With the portability system outlined above, most of the macros that you create will translate directly back and forth between the two spreadsheets.

LOTUS 1-2-3 AUTO-EXECUTING MACROS

To make an auto-executing macro for Lotus 1-2-3, just give the macro the special name '\0 (single-quote, backslash, zero). Assigning this name to a macro causes Lotus 1-2-3 to automatically execute the macro as soon as Lotus retrieves the spreadsheet.

The auto-executing macro for this example is shown in Figure 7-1. The example illustrates two methods to achieve portability. The first method is shown in cells BV8 and BV9. The {LET CALL_BASE,"{APP3}"} and {LET RET_BASE,"{ESC}"} commands change the contents of cells named CALL_BASE and RET_BASE, respectively. If you recall, the command sequences to access and return from the @BASE main menu are dependent on whether Lotus 1-2-3 or Symphony is being used. The CALL_BASE and RET_BASE cells are subroutine programs that access and return from the @BASE menu system using Release-specific commands. This method of direct modification is one way to achieve program portability and will be discussed in more detail later.

The auto-executing macro also places a number corresponding to the Lotus 1-2-3 Release into a cell called RELEASE. The RELEASE cell represents a second way to achieve portability. The RELEASE information will be used later in conjunction with {IF} commands to determine which set of menu commands to issue, thereby achieving program portability.

To determine the Lotus 1-2-3 Release, the macro takes advantage of differences in program features and defects that occurred in subsequent releases of Lotus 1-2-3. The first test is the /XGO~ test. This command sequence branches control to the range named O if the Release supports the /X macro key words. Lotus 1-2-3/G does not support the /X macro key words. With Lotus 1-2-3/G, the program interprets the slash and brings up the menu system. X is not a menu option, so Lotus 1-2-3/G selects the Graph menu, O selects Options, and the tilde selects Titles (i.e., the first option). Then the {ESC 4} command returns to the spreadsheet. If the macro does not branch to the O routine, the macro was executed by Lotus 1-2-3/G and the macro is aborted because @BASE does not support this Release of Lotus.

If /X causes a branch to O, further testing is performed to determine whether the Release is 2.0, 2.01, 2.2, or 3.X. The first test, based on @ISERR(1+"A"), tests for a program defect that occurred only in Release 2.0. In Release 2.0, calculations that mix numbers and strings generated an error. If the result of 1+"A" is ERR, the macro enters a 2.0 into the RELEASE cell.

Calculations and @functions return "ERR" if an error occurs in their arguments.

```
--------BU-------BV------BW------BX-------BY-------BZ-------CA-------CB-----
 1                          /*LIMS PROGRAM*/
 2           /*USES @BASE ADD-IN PROGRAM TO ARCHIVE AND RETRIEVE DATA*/
 3                  /*REVISED 5-DEC-90;  LOUIS M. MEZEI*/
 4
 5                /*AUTO-EXECUTING MACRO FOR LOTUS 1-2-3*/
 6             /*DETERMINE RELEASES /G, 2.0, 2.01, 2.2, AND 3.X*/
 7           /*QUIT IF RELEASE IS /G OR 3.X (NOT SUPPORTED BY @BASE)*/
 8  \0       {LET CALL_BASE,"{APP3}"} /*SPECIFY COMMAND FOR @BASE MENU*/
 9           {LET RET_BASE,"{ESC}"}   /*POST INPUT & RETURN FORM @BASE*/
10           /XGO~{ESC 4}    /*TEST WHETHER RELEASE IS 1-2-3/G*/
11           {QUIT}          /*1-2-3/G NOT SUPPORTED BY @BASE, QUIT*/
12
13             /*DETERMINE RELEASES 2, 2.01, 2.2, AND 3.X*/
14  0        {IF @ISERR(1+"A")}{LET RELEASE,2.0}{VU_OPEN}{MENUBRANCH MAIN_MU}
15           {LET RELEASE,2.01}      /*TENTATIVELY IDENTIFY RELEASE AS 2.01*/
16           {IF #NOT#@ISERR(@CELL("FILENAME",TST_CELL))}{LET RELEASE,2.2}
17           {IF #NOT#@ISERR(@CELL("COORD",TST_CELL))}{LET RELEASE,3}{QUIT}
18           {CALC}                  /*FORCE SPREADSHEET RECALCULATION*/
19           {VU_OPEN}               /*OPEN VIEW FILES*/
20           {MENUBRANCH MAIN_MU}    /*BRING UP MAIN MENU*/
21
22                /*AUTO-EXECUTING MACRO FOR SYMPHONY*/
23           /*DETERMINE RELEASES 1.0, 1.1, 1.2, 2.0, 2.2*/
24           /*QUIT IF RELEASE IS 1.0 OR 1.1 (NOT SUPPORTED BY @BASE)*/
25  AUTOEXEC {IF @ISERR(@VALUE("(1)"))}{LET RELEASE,4.10}{CALC}{QUIT}
26           {IF @ISERR(@PMT(-1,1,1))}{LET RELEASE,4.11}{CALC}{QUIT}
27           {LET RELEASE,4.12}      /*TENTATIVELY IDENTIFY AS RELEASE 1.2*/
28           {IF #NOT#@ISERR(@CELL("TYPE",TST_CELL))}{LET RELEASE,4.20}
29           {BLANK TST_CELL}{LET TST_CELL,4.22:string}/*TEST FOR REL 2.2*/
30           {IF @ISNUMBER(TST_CELL)}{LET RELEASE,4.22}
31           {BLANK TST_CELL}
32           {IF RELEASE=4.22}{LET CALL_BASE,"{TYPE}B/"}
33           {IF RELEASE=4.22}{LET RET_BASE,"FP{ESC}{TYPE}S"}{VU2_OPEN}
34           {IF RELEASE=4.12#OR#RELEASE=4.20}{LET CALL_BASE,"/@"}
35           {IF RELEASE=4.12#OR#RELEASE=4.20}{LET RET_BASE,"FP{ESC}"}{VU_OPEN}
36           {CALC}                  /*FORCE SPREADSHEET RECALCULATION*/
37           {MENUBRANCH MAIN_MU}    /*BRING UP MAIN MENU*/
38
39           /*OPEN VIEW FILES; CONFIGURE INDEX & JOIN SYSTEM*/
40  VU_OPEN  {IF @VIEWOPEN("BATCH.VU")=0}{BEEP}{QUIT}
41           {IF @VIEWOPEN("RESULTS.VU")=0}{BEEP}{QUIT}
42
43           /*OPEN SYMPHONY 2.2 VIEW FILES; CONFIGURE INDEX & JOIN SYSTEM*/
44  VU2_OPEN {IF @VIEWOPEN("BATCH.VU2")=0}{BEEP}{QUIT}
45           {IF @VIEWOPEN("RESULTS.VU2")=0}{BEEP}{QUIT}
46
47  TST_CELL                /*TEST CELL FOR RELEASE IDENTIFICATION*/
48  RELEASE                 /*LOTUS RELEASE:  2; 2.01; 2.2; 3*/
49                          /*SYMPHONY RELEASE:  4.1=REL 1.0; 4.11=REL 1.1*/
50                          /* 4.12=REL 1.2; 4.2 = REL 2.0; 4.22 = REL 2.2*/
```

Figure 7-1

The (1+"A") is ***nested*** within an @ISERR @function. @ISERR will translate an ERR returned from (1+"A") into a "true" (that is, a one). If an ERR is not returned by (1+"A"), then @ISERR will return a false (zero). Thus, the value returned from (1+"A") is translated into something that an {IF} command can understand.

If you recall from Chapter 6, an {IF} command allows a macro program to perform a specific task based on the result of a TRUE/FALSE test. If the result of the expression within the {IF} command does not have the numeric value of zero, Lotus considers the expression to be TRUE and the macro executes the instructions in the same cell, immediately following the {IF} command. If the result of the expression within the {IF} command has the numeric value zero, Lotus considers the expression to be FALSE and the macro continues down to the cell below the {IF} command.

Next, RELEASE is assigned the value 2.01. The remainder of the macro uses sequential logic to rule out all other Releases and, thereby, identify Release 2.01.

A "FILENAME" argument for the @CELL @function first appeared in Releases 2.2 and 3. The {IF} test in cell CW15 checks the validity of the "FILENAME" argument in the @CELL @function. If the formula does not generate an error, the macro places a 2.2 into the RELEASE cell. If an error is generated, Release 2.01 is confirmed.

To differentiate between Release 2.2 and 3.X, the program tests the "COORD" argument for the @CELL @function. This argument became available with Release 3. If the @CELL @function does not generate an error, the macro enters the value 3 into RELEASE and aborts because @BASE does not support Release 3 at this time.

{CALC} forces a recalculation of the spreadsheet so that the current value of RELEASE appears in the cell.

To create the example macro, start by typing the text as you see it in Figure 7-1. In addition to the program in Figure 7-1 you will need to create cells to accept values from the {LET} commands in the program. To create the cells, type CALL_BASE and RET_BASE into cells FG29 and FG31, respectively.

Before you can use any macro, you must assign the macro a formal name and tell Lotus 1-2-3 where the macro begins. Assigning a formal name to the macro will ***activate*** the macro. To activate the '\0 and O macros, move the cell-pointer to cell BU8, issue the /*R*ange *N*ame *L*abel *R*ight command, press period (.), use the [DOWN] arrow key to highlight cells BU8..BU14, and press [RETURN]. Repeat for cells FG29..FG31. Store the spreadsheet.

SYMPHONY AUTO-EXECUTING MACROS

The format of a Symphony auto-executing macro is also shown in Figure 7-1. The name of the macro is AUTOEXEC. AUTOEXEC is similar in format and rationale to '\0, but differs in that Symphony 2.2, receives processing that is different from earlier Releases of Symphony. If you recall from previous chapters, there are slight differences in the way Symphony 2.2, and previous Releases access and utilize @BASE. The following are some examples:

- Before the @BASE main menu can be accessed in Symphony 2.2, the environment must be changed to "BASE". Previous Releases are accessed directly from the SHEET environment.

- Symphony 2.2, has a different view file format and file extension name (.VU2) than previous Releases. Previous Symphony Releases had ".VU" file extensions and were fully compatible with Lotus 1-2-3-generated view files.
- The Option menu commands for Symphony 2.2, contain an extra ***F***ile menu choice before ***I***ndex options can be accessed.

These differences necessitate identification of the Release of Symphony being used. Having to deal with these details makes the AUTOEXEC macro a little more complicated than '\0. Let us examine some of the details in this macro.

AUTOEXEC begins by determining whether the Release is 1.0 or 1.1 using a pair of {IF} tests. The first {IF} test examines the way that @VALUE handles numbers in parentheses. In Release 1.0, @VALUE did not support the use of parentheses as an indication of negative numbers. The second {IF} test examines whether negative numbers are acceptable as the first argument of the @PMT @function. Negative numbers were not supported until Release 1.2. Therefore, if the @PMT @function evaluates to ERR, the program is Release 1.1. If either of these two {IF} tests evaluates to true, the macro aborts because @BASE does not support these Releases of Symphony.

Next, RELEASE is assigned the value 4.12, the code for Release 1.2. Dividing the Symphony Release number by ten and adding 4 separates the Symphony Releases from the Lotus 1-2-3 Releases. Later, when the RELEASE cell is tested using {IF} commands, this technique provides a degree of programming efficiency. For example, a single {IF} test can be used to determine whether the RELEASE cell is less than or greater than four and actions can proceed accordingly.

The remainder of the macro uses sequential logic to rule out all other Releases and thereby identify Release 1.2. The first test identifies Releases prior to 2.0. With prior Releases, an ERR is returned if the second argument of @CELL is not a range (e.g., the second argument is CW8 instead of CW48..CW48). If @CELL has a single cell address and no error results, the Release must be 2.0 or higher. That is, Release 1.2 is ruled out.

The lower-case "string" portion of the {LET} command in cell CW28 identifies Release 2.2 because of a defect in the way that the {LET} command operates in this Release. If all of the letters in "string" are lower-case, Release 2.2 interprets the characters in the {LET} command as a value instead. The @ISNUMBER @function in cell CW29 is used to check the value placed in TST_CELL. If TST_CELL contains a number, the program is Release 2.2.

After the Symphony Release is determined, four {IF} tests in cells BV32..BV35 determine the Release and place the appropriate command sequence to access the @BASE main menu into the cell called CALL_BASE and the sequence to return from the @BASE menu into a cell called RET_BASE. As explained in the previous section, the CALL_BASE and RET_BASE cells are subroutine programs that access and return from the @BASE menu system using Lotus program-specific commands. This method of direct modification is a second way to achieve program portability and will be discussed in more detail later.

You can create a similar macro program for your application by modifying the example macro for your particular needs. To create the example macro, start by typing the text as you see it in Figure 7-1. It is very important that there be at least one blank row between macros because a blank row signals the end of a macro to Lotus 1-2-3 and Symphony.

When you create your own auto-executing macro, the name can be a single quote-backslash ('\) followed by a letter. Alternatively, the name can be a string of text characters (without the backslash). When choosing a name for a macro, be as descriptive as possible. For this reason, I chose the name AUTOEXEC. (Naming the macro AUTOEXEC would still allow you to run the macro with the USER key (F7) anytime during your session.)

Before you can use the macros in Column BV, you must assign them formal names. These names tell Symphony where the macros begin. Assigning a formal name to a macro will ***activate*** the macro. Name a macro by issuing the /**R**ange **N**ame **L**abel **R**ight command. When prompted for the range of labels, press [ESC]ape, use the arrow keys to move to the cell containing the AUTOEXEC, type a period (.), and arrow down to highlight all of the cells with names in column BU. Press [RETURN].

In addition to the program in Figure 7-1 you will need cells named CALL_BASE and RET_BASE. These two cells accept values from the {LET} commands in the AUTOEXEC program. If you did not create the cell as part of the previous section, type CALL_BASE into cell FG29 and RET_BASE into cell FG31. Use the /**R**ange **N**ame **L**abel **R**ight command to name the cells.

Symphony needs to know more about the AUTOEXEC macro before it can use AUTOEXEC for auto-execution. Specifically, Symphony needs to know that the macro called AUTOEXEC is the one to run when the spreadsheet is retrieved. To achieve the flexibility that allows a user to specify names other than '\0 for auto-executing macros, Lotus requires you to place the name of an auto-executing macro into the spreadsheet's settings. Do so by issuing the [SERVICES] **S**ettings **A**uto-Execute **S**et command and typing AUTOEXEC. Store the spreadsheet. (NOTE: If the spreadsheet had been stored as a Lotus 1-2-3 file and is now being used as a Symphony spreadsheet, the Auto-Execute specification would be lost and you would need to re-set the specification.)

PROGRAM PORTABILITY

As stated previously, Lotus spreadsheets have minor differences in the way that some of the menu commands are issued; but the macro commands and @functions are identical. Therefore, if you can detect which Lotus spreadsheet is using a macro program, you can issue the appropriate menu commands. {LET} commands in each of the two auto-executing macros of Figure 7-1 help you with these tasks.

After the Release of Lotus 1-2-3 or Symphony is determined, a {LET RELEASE} command places a Release-encoded number into the cell named RELEASE. By testing for the encoded number in the RELEASE cell with {IF} commands, the program decides which macro subroutines to execute or which sets of menu commands to issue. For example, you could use the following lines in a program to issue the @BASE menu command to select a primary index for an open database file:

```
      :
{CALL_BASE}
{IF RELEASE<>4.22}OIS
{IF RELEASE=4.22}/OIFS
      :
```

The {LET} commands in cells BV8, BV9, and BV32..BV35 achieve the same goals. However, they change the program directly. That is, the auto-executing macro can customize cells that lie in the path of macro execution. After the initialization macro completes execution, these cells contain spreadsheet-specific commands that execute when encountered by the program. In this example, they change the contents of the cells named CALL_BASE and RET_BASE.

SYSTEM INITIALIZATION

The VU_OPEN and VU2_OPEN subroutines illustrate a common form of system initialization. Because of differences in view file formats, the VU2_OPEN subroutine is used for Symphony 2.2, and VU_OPEN is used for all other Releases.

Both subroutines use the @VIEWOPEN() @function to open view files. After each view file is opened, the instructions in the view file are executed, and the primary, all lookup, and all index files specified in the view instructions are thereby opened. The virtual fields (joined and computed) are also activated. In short, the subroutines automate the process of initializing the entire extended relational database system.

The @VIEWOPEN @functions have been placed in {IF} commands so that the returned values can be tested. @VIEWOPEN returns a 1 if the view file was successfully opened, otherwise it returns a 0. If a 0 is returned, the speaker is sounded to announce that an error has occurred and the macro aborts.

STARTING MENUS FROM AUTO-EXECUTING MACROS

A feature that you will want to include in your program is one that allows users to retrieve a spreadsheet and be automatically presented with a menu of all functional options available in a program. A program that automatically takes care of all details for the user without the user having knowledge of them is user friendly.

The following discussion describes how this system works: After RELEASE has been updated, the program lines have been modified, and the view files opened, the {MENUBRANCH MAIN_MU} command displays a custom menu called MAIN_MU in the control panel. A custom menu looks and works just like the ones that Lotus 1-2-3 or Symphony presents to you when you press the / key.

Each menu option has its own capsule description (help line) that is displayed when the menu option is highlighted with the menu-pointer. Each menu option is also tied to a macro routine. When a user selects a menu option, the corresponding macro routine

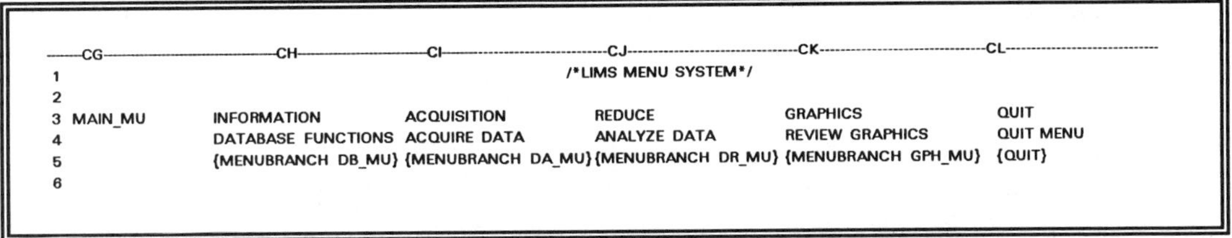

	CG	CH	CI	CJ	CK	CL
1				/*LIMS MENU SYSTEM*/		
2						
3	MAIN_MU	INFORMATION	ACQUISITION	REDUCE	GRAPHICS	QUIT
4		DATABASE FUNCTIONS	ACQUIRE DATA	ANALYZE DATA	REVIEW GRAPHICS	QUIT MENU
5		{MENUBRANCH DB_MU}	{MENUBRANCH DA_MU}	{MENUBRANCH DR_MU}	{MENUBRANCH GPH_MU}	{QUIT}
6						

Figure 7-2

executes.

You can execute a menu in one of two ways. The first way is to use the {MENUCALL menu_name} command; the second is to use the {MENUBRANCH menu_name} command. These two commands are similar. They both pause macro execution, go to a specified location, and display the menu at the location.

However, each command treats a menu differently. {MENUCALL} treats a menu like a subroutine. That is, it returns to the original macro when the subroutine instructions in the menu have been completed. With {MENUBRANCH}, the macro "jumps" to the menu. Execution terminates in the original macro and total control transfers to the menu. Then, the menu executes. When the last instruction in the menu is encountered, macro execution will terminate (unless another branching command is encountered).

As you can see, a significant difference exists between the two menu commands. This dissimilarity can be important because Lotus limits the number of subroutine levels that you can call from within a program. This concept was introduced for macros in Chapter 6. Let us review this concept and apply the concept to menus.

If a macro calls another macro as a subroutine, Lotus is supporting two ***levels*** of subroutines. Likewise, if a second subroutine calls a third subroutine, Lotus is supporting three levels, and so on. The number of subroutine levels that either Lotus 1-2-3 or Symphony can support (without returning to the first macro level) is 31.

If you exceed the subroutine limit by calling a macro from itself and/or by trying to ***nest*** too many subroutines, the macro aborts and an error message appears. Because a {MENUCALL} acts just like a subroutine call, the same problem would occur if you tried to repeatedly re-use a menu from a {MENUCALL}.

MENUS

Figure 7-2 is an example of a menu. Note that this figure has an expansion of column widths to expose all text. MAIN_MU is a simple menu. Take a moment to examine this menu so that you will know how to construct a menu for your own applications.

All custom menus are structured with the same basic format. The first row of cells defines the options that will appear when the menu is displayed. The text for the first option must be in the cell specified in the {MENUBRANCH} or {MENUCALL} command. The text for the other options in the menu must be in the same row and in the columns immediately to the right of the column that contains the first option. If you want to be able to select a menu option by pressing the first letter of a prompt, you must designate the first letter of each option as a unique upper-case letter, digit, or symbol from the keyboard.

Option prompts can be of variable length. Symphony and Lotus 1-2-3 automatically place two empty spaces between prompts. The length of the prompts plus spaces must not exceed 80 characters. If the total length exceeds 80 characters, the spreadsheet truncates all entries longer than a specific length; the length depending on the number of prompts used.

Your menu can have a maximum of eight selections. If you try to use more than eight selections, the spreadsheet will ignore them. If you have fewer than eight selections, the cell to the right of the last menu choice must be empty. An empty cell signals the end of a menu list to Lotus.

The row immediately below the options contains the prompt's help message. The menu displays a help message when an option is highlighted. The longest possible help message is 75 characters. The spreadsheet will ignore all additional characters when a message is displayed.

The cells below a help message contain instructions that the menu choice should execute if a user makes that particular choice. These cells may contain macro commands, a subroutine call, a branch to a macro, a branch to another menu, or a branch to the same menu.

"LINKED" MENUS

Figure 7-3 shows a "tree" of menus that administer the database tasks of MAIN_MU. Each of these menus has a format that is exactly the same as the one described in the previous section. However, the menus are different in that they are "linked" together. That is, when a user makes a choice in one menu, a {MENUBRANCH} or {MENUCALL} command jumps to the next menu. This process continues down the hierarchy of menus until the user "zeros" in on the task that needs to be performed, at which time the appropriate macro is called upon to do the work. Figure 7-4 is a flow diagram illustrating the procession down the hierarchy of menus. (To simplify the flow diagram, the links between each submenu and the FN_MU have been omitted.)

For example, suppose a user wanted to print a Certificate of Analysis for a sample. The following sequence of events would occur:

- Choosing the Information option on the MAIN_MU branches to DB_MU (the database menu).
- Choosing Samples on the DB_MU branches to the SR_MU (the samples/results menu).
- Choosing Certificate on the SR_MU menu first sets the FORM definition to 3 and then calls the TYPE_MU as a subroutine, which allows the type of assay to be selected.

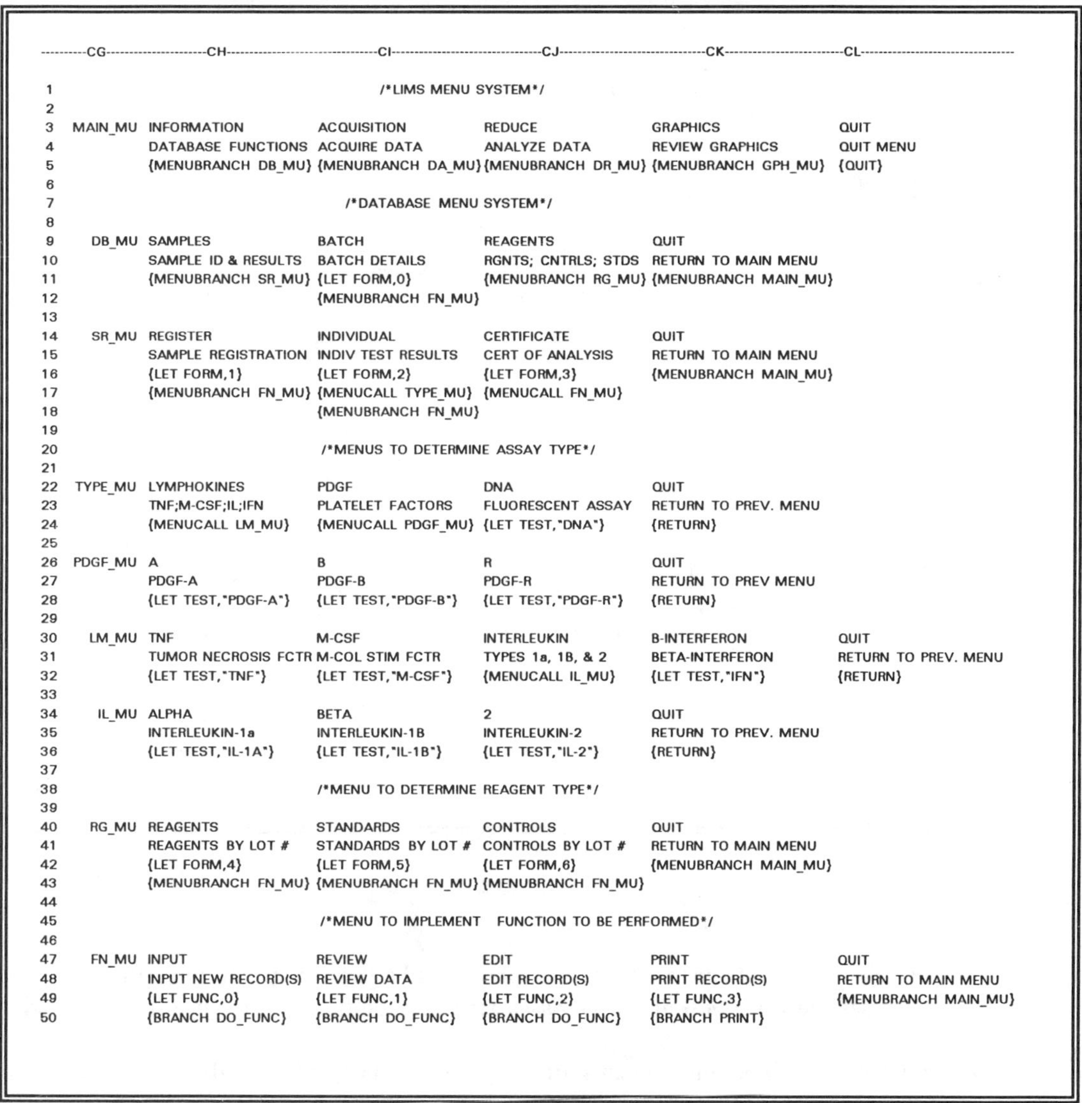

	CG	CH	CI	CJ	CK	CL
1			/*LIMS MENU SYSTEM*/			
2						
3	MAIN_MU	INFORMATION	ACQUISITION	REDUCE	GRAPHICS	QUIT
4		DATABASE FUNCTIONS	ACQUIRE DATA	ANALYZE DATA	REVIEW GRAPHICS	QUIT MENU
5		{MENUBRANCH DB_MU}	{MENUBRANCH DA_MU}	{MENUBRANCH DR_MU}	{MENUBRANCH GPH_MU}	{QUIT}
6						
7			/*DATABASE MENU SYSTEM*/			
8						
9	DB_MU	SAMPLES	BATCH	REAGENTS	QUIT	
10		SAMPLE ID & RESULTS	BATCH DETAILS	RGNTS; CNTRLS; STDS	RETURN TO MAIN MENU	
11		{MENUBRANCH SR_MU}	{LET FORM,0}	{MENUBRANCH RG_MU}	{MENUBRANCH MAIN_MU}	
12			{MENUBRANCH FN_MU}			
13						
14	SR_MU	REGISTER	INDIVIDUAL	CERTIFICATE	QUIT	
15		SAMPLE REGISTRATION	INDIV TEST RESULTS	CERT OF ANALYSIS	RETURN TO MAIN MENU	
16		{LET FORM,1}	{LET FORM,2}	{LET FORM,3}	{MENUBRANCH MAIN_MU}	
17		{MENUBRANCH FN_MU}	{MENUCALL TYPE_MU}	{MENUCALL FN_MU}		
18			{MENUBRANCH FN_MU}			
19						
20			/*MENUS TO DETERMINE ASSAY TYPE*/			
21						
22	TYPE_MU	LYMPHOKINES	PDGF	DNA	QUIT	
23		TNF;M-CSF;IL;IFN	PLATELET FACTORS	FLUORESCENT ASSAY	RETURN TO PREV. MENU	
24		{MENUCALL LM_MU}	{MENUCALL PDGF_MU}	{LET TEST,"DNA"}	{RETURN}	
25						
26	PDGF_MU	A	B	R	QUIT	
27		PDGF-A	PDGF-B	PDGF-R	RETURN TO PREV MENU	
28		{LET TEST,"PDGF-A"}	{LET TEST,"PDGF-B"}	{LET TEST,"PDGF-R"}	{RETURN}	
29						
30	LM_MU	TNF	M-CSF	INTERLEUKIN	B-INTERFERON	QUIT
31		TUMOR NECROSIS FCTR	M-COL STIM FCTR	TYPES 1a, 1B, & 2	BETA-INTERFERON	RETURN TO PREV. MENU
32		{LET TEST,"TNF"}	{LET TEST,"M-CSF"}	{MENUCALL IL_MU}	{LET TEST,"IFN"}	{RETURN}
33						
34	IL_MU	ALPHA	BETA	2	QUIT	
35		INTERLEUKIN-1a	INTERLEUKIN-1B	INTERLEUKIN-2	RETURN TO PREV. MENU	
36		{LET TEST,"IL-1A"}	{LET TEST,"IL-1B"}	{LET TEST,"IL-2"}	{RETURN}	
37						
38			/*MENU TO DETERMINE REAGENT TYPE*/			
39						
40	RG_MU	REAGENTS	STANDARDS	CONTROLS	QUIT	
41		REAGENTS BY LOT #	STANDARDS BY LOT #	CONTROLS BY LOT #	RETURN TO MAIN MENU	
42		{LET FORM,4}	{LET FORM,5}	{LET FORM,6}	{MENUBRANCH MAIN_MU}	
43		{MENUBRANCH FN_MU}	{MENUBRANCH FN_MU}	{MENUBRANCH FN_MU}		
44						
45			/*MENU TO IMPLEMENT FUNCTION TO BE PERFORMED*/			
46						
47	FN_MU	INPUT	REVIEW	EDIT	PRINT	QUIT
48		INPUT NEW RECORD(S)	REVIEW DATA	EDIT RECORD(S)	PRINT RECORD(S)	RETURN TO MAIN MENU
49		{LET FUNC,0}	{LET FUNC,1}	{LET FUNC,2}	{LET FUNC,3}	{MENUBRANCH MAIN_MU}
50		{BRANCH DO_FUNC}	{BRANCH DO_FUNC}	{BRANCH DO_FUNC}	{BRANCH PRINT}	

Figure 7-3

- If DNA is selected from the TYPE_MU, the type of assay is assigned and the program returns to SR_MU. If one of the other two choices are made, either LM_MU or PDGF_MU are presented. If Interleukin is selected on the LM_MU, IL_MU is presented. Based on the final selection, the type of assay is assigned and the program returns to SR_MU.
- Upon returning to SR_MU, a branch is made to FN_MU.

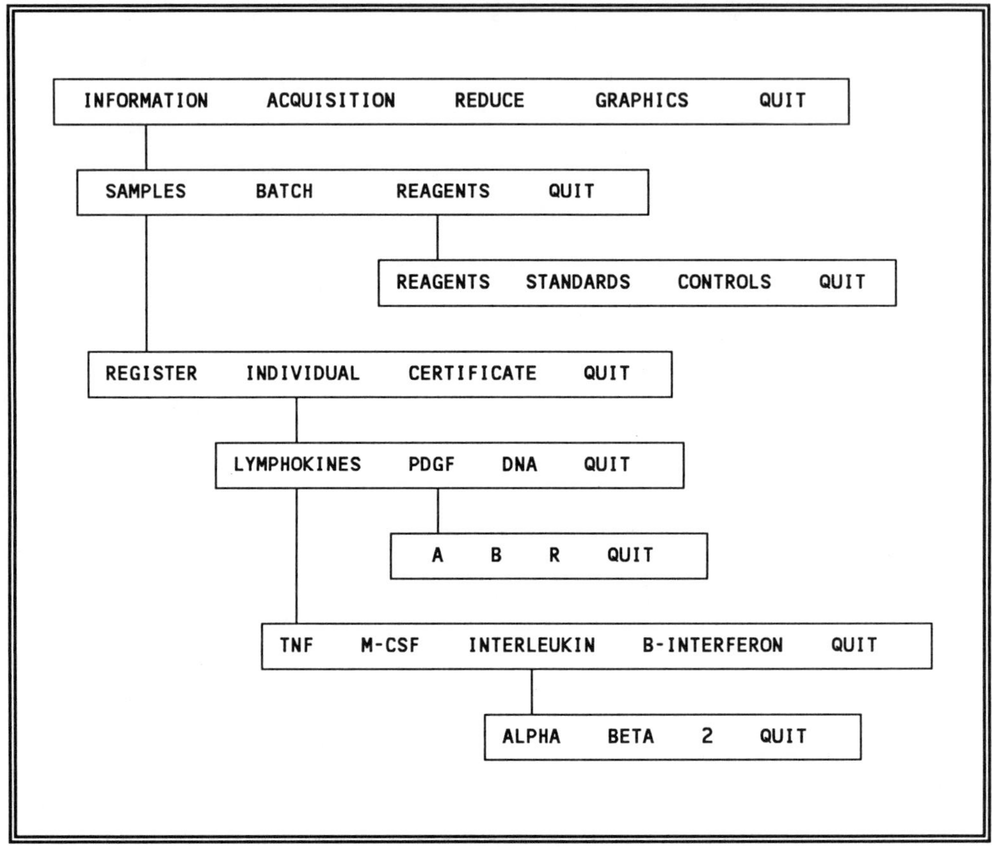

Figure 7-4

- FN_MU contains the Print menu choice and a generic subroutine ({PRINT}) that uses the form definition and assay type as specifications for printing the report.

As you can see, hierarchical menus are very user friendly because they let a user know what the program can do and what options are available at each point in the program. A list of descriptions for each menu acronym is provided in Figure 7-5. The following are some programming features of the menu tree shown in Figure 7-3:

- All menu choices within a menu have first letters that are unique.
- All menus have less than eight menu choices.
- Menu choices within menus have been kept to a minimum to decrease clutter and provide for future expansion.
- There is at least one open row between menus. There is also an open column

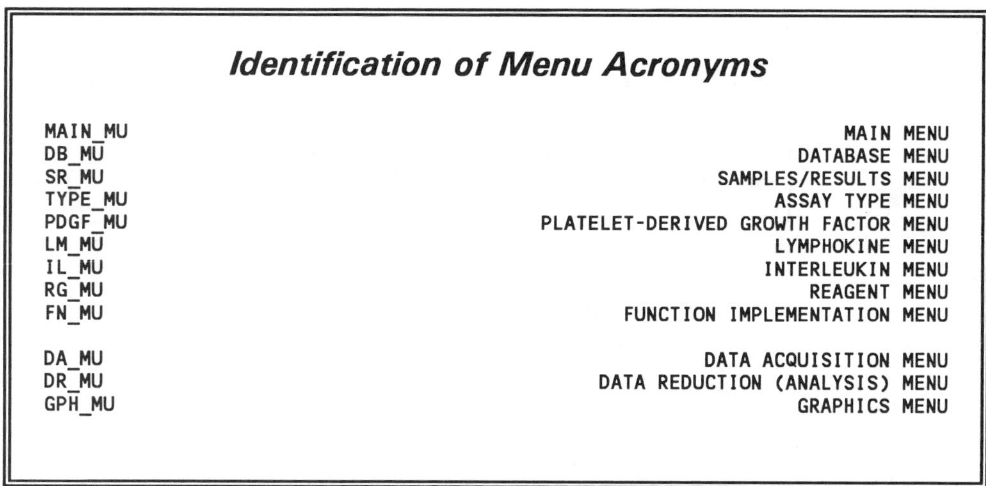

Identification of Menu Acronyms

MAIN_MU	MAIN MENU
DB_MU	DATABASE MENU
SR_MU	SAMPLES/RESULTS MENU
TYPE_MU	ASSAY TYPE MENU
PDGF_MU	PLATELET-DERIVED GROWTH FACTOR MENU
LM_MU	LYMPHOKINE MENU
IL_MU	INTERLEUKIN MENU
RG_MU	REAGENT MENU
FN_MU	FUNCTION IMPLEMENTATION MENU
DA_MU	DATA ACQUISITION MENU
DR_MU	DATA REDUCTION (ANALYSIS) MENU
GPH_MU	GRAPHICS MENU

Figure 7-5

cell to the right of the first row of each menu.

- The menu tree has been organized to collect related items into submenus and make the submenus accessible from single menu choices.
- The menus have been organized in the spreadsheet to correspond to the hierarchy in which they are called. This organization allows easier review during planning and programming phases.
- A cell called "FORM" is used throughout the menu system. FORM holds an encoded value that specifies the report that needs to be generated. When the downstream program generates a report, it will use the value in the FORM cell to determine which type of report is to be generated, form to use (i.e., Chapter 3 template), database to use, and which subroutines to use when preparing the report.
- A cell called "TEST" is also used. The TEST cell contains a value that codes for assay type. By providing menu choices instead of user input, a reliable test code is obtained and circumvents the problem of different users assigning different names to the same test.
- {MENUBRANCH} is used when moving between major menus and when the path after the menu has completed its task is different from the path calling the menu.
- {MENUBRANCH} is also used when menus need to be called repeatedly during a session. {MENUBRANCH} allows unlimited menu branching.
- The {MENUBRANCH FN_MU} at the end of menu choices in FN_MU is a common way to repeatedly return the user back to the same menu. The {MENUBRANCH} allows unlimited use of the menu. For analogous reasons, a {MENUBRANCH MAIN_MU} is used under the Quit option.

- {MENUCALL} is used when menus need to be generic and the path back to the original calling program or menu is not known. For example, the TYPE_MU can be used by several programs. By using {MENUCALL} within the hierarchy, each menu choice will return to the next higher menu when the menu's tasks have been completed. When TYPE_MU is finally reached, program control will revert back to the calling macro or menu. Thus, a program in the Acquisition branch of the tree can also call TYPE_MU, gain user input on the type of test to be performed, and issue pertinent instrument commands.
- If a {MENUCALL} is used to call a menu, a {MENUBRANCH} or {BRANCH} command does not appear in the menu. For example, the IL_MU does not branch to lower menus, it calls them as subroutines. If lower menus were branched to, the program flow would be broken, and a return to the LM_MU would not be possible.
- The {RETURN} command is optional. Program control will revert back to the calling program when the first open row is encountered. The {RETURN} command is used under Quit options to make the flow of a program easier to follow. If a menu has been accessed by a {MENUCALL}, the program would terminate when either {RETURN} or an open row was encountered.
- The use of Lotus key words was avoided. For this reason TEST was used instead of TYPE, because the latter is a Lotus key word.
- Programming within menus has been kept to a minimum.
- The menu tree serves as a flow diagram for the system and pinpoints the macro programs that need to be developed.
- The menu system follows the structure and database specifications defined in Chapter 2 and the report templates defined in Chapter 3. For example, the DB_MU, SR_MU, and RG_MU menu choices collectively correspond to the database files and forms defined in Chapters 2 and 3. The TYPE_MU and related submenus correspond to the Types of Assays list in Chapter 2.

It would be instructive to trace each pathway in the menu system in Figure 7-3 from the top menu down to the bottom menu to get a feel for the structure of the menu tree. You should try to understand how the tree acts like a funnel, narrowing down information as each subsequent menu is encountered. The following are some major pathways through the menu system:

- Information → Samples → Assign FORM → Assign TYPE → Implement.
- Information → Batch → Implement.
- Information → Reagents → Assign FORM → Implement.

To add more functionality to a menu system, you could add more branches to the tree. For example, to add the capability to issue End-Of-Month, Sample Status, Overdue Sample, Backlog, Reagent Utilization, Instrument Utilization, Cost Per Test, etc., reports, you could add a "Management" menu choice to the DB_MU. The Management menu

```
-------CG------------CH---------------CI-------------CJ----------------
54
55                /*TEMPORARY PROGRAM TOOLS TO ALLOW MENU SYSTEM TO OPERATE*/
56       FUNC
57       FORM
58       TEST
59
60    DO_FUNC     {MENUBRANCH FN_MU}
61
62      PRINT     {MENUBRANCH FN_MU}
63
64         \M     {MENUBRANCH MAIN_MU}
65
```

Figure 7-6

choice would then lead to a MG_MU containing menu choices to pinpoint the information needed. The Management menu should be similar in design to the SR_MU system in that the management reports should be separated according to category, each category being given a menu choice, and each menu choice leading to lower menus.

BRANCHING FROM MENUS

The "Input," "Review," and "Edit" menu choices in FN_MU show how to execute a macro from within a menu. These choices use the {BRANCH} command to transfer program control to {DO_FUNC}. That is, program execution terminates in the menu and total control transfers to {DO_FUNC}.

If you use the {BRANCH} command to execute a macro you must have a {MENUBRANCH} command at the end of the macro. This command is needed to return to FN_MU. If {MENUBRANCH} were omitted, the program would terminate after the macro was completed.

QUITTING MENUS

Always include a Quit menu choice at the end of every menu. This choice allows a user to abort the menu's execution and either return to a previous menu or to the spreadsheet. By returning to previous menus, the user can proceed down menus without having to worry about ending up in a blind alley.

Also, one of the most important reasons for using a spreadsheet is that it allows a user to explore, manipulate, graph, perform statistics, and perform unique operations on data. A Quit menu choice allows a user to utilize these capabilities. The {QUIT} command that appears at the end of the example main menu terminates all macro and menu execution and returns the user to the spreadsheet.

```
---------------CR--------CS----------CT-------CU----------CV--------CW----
 1
 2
 3                          RANGE NAME TABLE
 4
 5        ASSAY_TYPE B4                          PDGF_MU       CH26
 6        AUTOEXEC   BV25                        PRINT         CH62
 7        BATCH      D4                          RAW_STD       B9..E14
 8        BCH_FM     AV1..BB22                   RECNO         A4
 9        BCH_HDR    AV25..BB25                  RELEASE       BV48
10        BCH_TAB    AV26..BB106                 RET_BASE      FH31
11        BC_CT      B39..B999                   RG_FM         BF25..BP45
12        B_SD       AV25..BB25                  RG_MU         CH40
13        CALL_BASE  FH29                        RUN_DT        C4
14        CERT_FM    Y1..AF15                    RUN_TECH      E4
15        CERT_HDR   Y19..AF19                   SID_FM        01..U19
16        CERT_TAB   Y20..AF30                   SLOPE         E32
17        DB_MU      CH9                         SRD_FM        AK1..AR25
18        DNA_TRIO   H39..H50                    SR_MU         CH14
19        DO_FUNC    CH60                        STATS         C25
20        FL_AVG     A21.G22                     STAT_TB       C25..F33
21        FL_CONC    A17..G18                    STDF          F9..F14
22        FL_REC     A3..G4                      STDX          A9..A14
23        FL_REP     B38..J50                    STD_FM        BF1..BP20
24        FL_TRIO    B4..D4                      TARGET        H39..J499
25        FN_MU      CH47                        TECH_COMS     B5
26        FORM       CH57                        TEMPLATE      A1..I51
27        FUNC       CH56                        TEST          CH58
28        IL_MU      CH34                        TST_CELL      BV47
29        INTERCEPT  F26                         TYPE_MU       CH22
30        LM_MU      CH30                        VU2_OPEN      BV44
31        LOT_RGNT   G4                          VU_OPEN       BV40
32        LOT_STD    F4                          \0            BV8
33        MAIN_MU    CH3                         \M            CH64
```

Figure 7-7

CREATING THE EXAMPLE MENU SYSTEM

To create the example menu system, type the text of Figure 7-3 into the template spreadsheet created in Chapter 3. Activate the menus using the /*R*ange *N*ame *L*abel *R*ight command and specifying CG3..CG47 as the range. In addition to the named ranges for the menu system, there are some program tools that need to be created. These temporary ranges will allow the menu to be explored and evaluated before actual programs and scratch-pad cells have been created. To create the temporary program tools, type the text shown in Figure 7-6 into the specified cells and use the /*R*ange *N*ame *L*abel *R*ight command to assign temporary range names to the cells in Column CG.

Figure 7-7 is a table that lists all of the named ranges created up to this point. The table includes ranges from templates created in Chapter 3 and macros and menus created in this chapter. The table was created using the /*R*ange *N*ame *T*able command. You should issue this command, specify cell CR5 as the starting cell for the table, and compare your table with Figure 7-7. This comparison will show if you have missed or incorrectly identified any named ranges. This check is important because accuracy in naming ranges is an absolute requirement for a program to execute properly. If you find any discrepancies, resolve them.

For example, if you missed creating a named range that was used in a calculation within a {LET} or {PUT} command, the command would place the name into the specified cell, not the value of the named range cell. Having the table handy while you are programming will also make existing range names available so that you do not try to re-use names.

To activate the menu system, hold down the [ALT] key and press M.

SUMMARY OF WHAT HAS BEEN ACCOMPLISHED

This chapter explained how to create auto-executing macros that

- determine the Release of Lotus software being used,
- modify cells (based on Lotus Release) to achieve program portability, and
- activate menu systems.

The importance of well-designed menus was also discussed. A menu system coordinates all of the functionality of your LIM system. Therefore, menus are key to efficient programming, reliable program performance, and user friendly operation.

WHAT'S NEXT?

The next chapter will explain how create a library of generic subroutine utilities that can be combined to form fully functional LIM system database management programs.

8

Generic Subroutine Utilities

Each macro program that supplies functionality to the menu system created in Chapter 7 typically contains the following sections:

- User input to obtain specifications for the record(s) to be imported or exported.
- User input to enter or edit data within a record.
- Formatting of data to conform to the specifications used in database definitions.
- Selection of the primary index file(s) to use.
- Determination of existing record numbers or creation of new records.
- If data are exported, retrieval of data from a template and transfer to a database.
- If data are imported, retrieval of record number-specific data from a database and transfer of data to a template.

Thus, there will be a lot of common functionality to your programs. This common functionality is best supported by creating generic subroutine utilities and using the subroutines as building blocks within macro programs tailored to specific tasks to be completed. This subroutine approach to programming leads to programs that

- are easier to read and follow.
- contain easier-to-fix errors.
- are smaller.
- are less convoluted.
- are more efficient.
- use less memory.

This chapter describes a number of subroutines that will make your programming chores easier. The subroutines are generic because

- they cater to a wide variety of calling programs without depending on the circumstances under which they are called.
- they typically place values in general purpose buffer cells so that the information is available to any macro program.
- after they have completed their tasks, they effect a smooth transition back to the main program.

The subroutines are utilities because they perform specific functions in support of other programs. In this context, the subroutine utilities provide the functionality for the building blocks that are needed to create programs containing the sections listed above. Some of the subroutines contained in this chapter are comprised of a single {LET} command implementing an @BASE @DB or @NDX function. These @functions can be used directly in your programs. However, I have found it easier to create reliable programs using the subroutine approach. I also benefit from the uniformity of function that is provided by using the same subroutine throughout a program. Therefore, this approach is the one presented in this book.

PASSING ARGUMENTS TO SUBROUTINES

Before you can get a full understanding of the functionality of the subroutines in this chapter and how to use them, you need to know how to pass arguments to subroutines and how @BASE @functions expect the arguments to be formatted. Incorrectly formatted arguments are the most common cause for subroutine and @BASE @function failure. This section describes how to pass arguments to subroutines. The next section describes how to use arguments in @BASE @functions.

The {PAD} subroutine in Figure 8-1 is an illustration of the passing of arguments to subroutines. With most Lotus Command Language statements, you specify a keyword followed by one or more arguments. For example, {LET BUFFERA,1234} assigns the value 1234 to a cell called BUFFERA. In a similar manner, you can give values to macro subroutines that you create. When you pass arguments to a subroutine, you are giving values to the subroutine. In the {PAD} example, you call the subroutine by enclosing both the subroutine name and arguments within braces. For example,

```
{PAD "BATCH","ASSAY_TYPE",@INDEX(FL_REC,2,1)}
```

This statement is a subroutine call in which three arguments are passed. For each argument that you pass to a subroutine, you must also specify a cell to store the value being passed and how the arguments being passed are to be interpreted. This designation is accomplished with a DEFINE statement at the beginning of the subroutine. For example, the {PAD} subroutine begins with the following statement:

```
{DEFINE DB_FILE,FLD_NM,BUFFERA:VALUE}
```

When Lotus begins executing {PAD}, it processes the {DEFINE} command and stores the values passed to the {PAD} subroutine in cells named DB_FILE, FLD_NM, and BUFFERA, respectively. The ":VALUE" portion of the statement represents a significant difference in the way in which subroutine calls and Lotus macro commands and @functions operate. With Lotus macro commands and @functions, you can supply cell names, formulae, or actual values as arguments. With subroutine calls, you must supply the type of argument specified in the DEFINE statement. In general, the ":VALUE" suffix instructs Lotus to evaluate the argument before storing the value. That is, Lotus considers the argument to be a number, a cell address, or a formula. If you omit the ":VALUE" suffix, Lotus stores the argument as a label, exactly as it appears in the subroutine call. For example, if the ":VALUE" suffix had been omitted from the DEFINE statement in {PAD}, the string "@INDEX(FL_REC,2,1)" would have been stored in BUFFERA.

Conversely, if you had used the ":VALUE" suffix and had tried to pass a string to the subroutine, an error would have occurred. For example, if you had tried to pass "DNA" to {PAD} using the following statement, the macro would have aborted and an error message would have been issued:

```
{PAD "BATCH","ASSAY_TYPE","DNA"}
```

To circumvent this problem, when passing strings, you can place the strings in token @functions to fulfill the number, cell address, or formula requirement. @UPPER, @TRIM, and @CLEAN @functions make no substantive changes on strings containing no spaces. @UPPER is independent of spaces. To create a functionally reliable statement to call {PAD} using "DNA", the following could be used:

```
{PAD "BATCH","ASSAY_TYPE",@UPPER("DNA")}
```

The choice of whether to use the ":VALUE" suffix depends on the intended use of a subroutine and argument. For example, if a subroutine will be called repeatedly during execution of {FOR} loops and the argument will change on each execution, you should use the ":VALUE" suffix. If, on the other hand, the value of the argument will not change, do not use the ":VALUE" suffix. Arguments that usually do not change are index and database file names.

The number of arguments passed to a subroutine must match the number of arguments in the DEFINE statement. Otherwise, Lotus issues an error message. If you omit the DEFINE statement, no values are transferred, meaning that the subroutine cannot use any of the arguments you specified in the subroutine call.

@BASE @FUNCTIONS

Most of the subroutines presented in this chapter are based on @BASE @DB and @NDX @functions. @BASE @functions are specialized formulae that let you work with database files from within a Lotus spreadsheet. The @BASE Option Pac @NDX @functions communicate with index file data to establish efficient connections between the spreadsheet and database. @NDX @functions are primarily used to locate record numbers. Once record numbers are located, @DB @functions enable you to analyze, perform calculations, and update database files from within a spreadsheet while the database files remain on disk.

Typically, @BASE @functions are nested within Lotus {LET} commands and are used in macro programs. {LET} commands provide a way to calculate individual @BASE @functions without causing an entire spreadsheet to recalculate. By calculating @BASE @functions on demand, operation is efficient and performed in an orderly fashion. You can use @BASE @functions to perform the following types of @BASE operations:

- Open and close database and index files.
- Locate records and retrieve record numbers.
- Add new records to a database.
- Update existing database records.
- Retrieve information from fields in database records.
- Perform statistical analyses of data in a database file.

Some @BASE @functions duplicate tasks you can also perform with @BASE menu commands. In general, it is better to use an @BASE @function rather than issue the corresponding menu command sequence for the following reasons:

- Using an @BASE @function is usually faster.
- When a macro cannot execute a command sequence successfully, the macro terminates and issues an error message. This is not user friendly. However, when a macro cannot execute an @BASE @function successfully, the @BASE function returns an error value (ERR or 0) instead of causing the macro to terminate. When you use @BASE @functions within {LET} commands, you can follow the {LET} commands with {IF} commands to check whether the @function executed successfully. This method allows you to test for errors and handle them accordingly.

@BASE @functions are similar in format to Lotus @functions and return single values. The general format of an @BASE @function is:

```
@DBfunction name(argument1,argument2,...)
@NDXfunction name(argument1,argument2,...)
```

The ***function name*** tells @BASE which @DB or @NDX function to perform. The ***arguments*** are the parameters used in the function's operations or calculations. The first argument is usually the alias of the database file you want @BASE to perform the @function on. The remaining arguments can be field names, record numbers, or values. Following are some guidelines for arguments:

- Do not include spaces in @BASE @functions.
- Arguments which refer to strings must be enclosed in single or double quotes.
- Do not use quotes for cell references, range references, and numbers.
- Arguments must be separated by commas.

The following terms represent the parameters used in @BASE @function arguments:

- ALIAS is the internal name @BASE uses to identify an open database file.
- DATABASE is the full name of a database file, including the extension.
- FIELD NAME is the name of a field in a specified database.
- VALUE is the value in a particular field.
- RECNO is a record number in a specified database file.

USER INTERFACE SUBROUTINES

Figure 8-1 shows subroutines that obtain user input and format the input to coincide with database definitions. The following subsections describe the function, use, and programming features of each subroutine.

{USR_INP}

FUNCTION: {USR_INP} allows a user to input values directly into cells of a template until [CONTROL] and [BREAK] are pressed.

INPUT: {USR_INP} requires a subroutine name in a cell named SUB_PTR.

OUTPUT: None.

EXAMPLE:

```
{LET SUB_PTR,"RGT_EXP"}
{BRANCH USR_INP}
```

NOTES ON USAGE: {USR_INP} illustrates a way to allow user input into a template. The {?} command in cell FH6 allows all operations: Input, [F2]-Edit, Lotus menu commands, etc. Alternatively, you could build a system using {GETLABEL} and {GETNUMBER} commands. However, this second system is less flexible and tends to require quite a large program to implement. Routine applications do not usually warrant the extra programming.

PROGRAM DESCRIPTION: {USR_INP} features some important programming tools. The Lotus {ONERROR} command establishes a procedure to be followed if an error occurs during macro execution. A Lotus 1-2-3 or Symphony error is any action that causes a beep, a macro to abort, and an error message to be displayed in the lower left corner of the display. Typically, such an error occurs when a spreadsheet cannot successfully complete a command with the specifications provided. When the [CONTROL] and [BREAK] keys are pressed together to abort a macro program, the termination is actually handled as an error condition. The syntax of the {ONERROR} command is

```
--------FG--------FH-------FI-------FJ-------FK-------FL-------FM-------FN---
 1                    /*GENERIC SUBROUTINE UTILITIES*/
 2
 3              /*KEEP ALLOWING INPUT UNTIL [CTRL][BREAK]*/
 4              /*NEEDS A SUBROUTINE NAME IN SUB_PTR CELL TO RETURN*/
 5  USR_INP     {ONERROR FIN_RGT}          /*SET ERROR BRANCH FOR LOOPING*/
 6  RGT_LP      {?}~{BRANCH RGT_LP}        /*LOOP TO ALLOW DATA INPUT*/
 7  FIN_RGT     {ONERROR ABORT}            /*RE-DIRECT ERROR TRAP*/
 8              {WINDOWSOFF}{PANELOFF}{CALC}        /*FREEZE THE DISPLAY*/
 9              {DISPATCH SUB_PTR}/*BRANCH TO SUBROUTINE SPECIFIED IN SUB_PTR*/
10
11              /*SEE IF A CHARACTER STRING; IF NOT, CONVERT IT TO ONE*/
12  MK_STRNG    {DEFINE BUFFERA:VALUE}
13              {IF @ISERR(@STRING(BUFFERA,0))}{RETURN}
14              {LET BUFFERA,@STRING(BUFFERA,0)}
15
16              /*SEE IF DATE IS ALREADY CONVERTED; IF NOT, CONVERT*/
17  MK_DT       {DEFINE BUFFERA:VALUE}
18              {IF @ISERR(@DATEVALUE(BUFFERA))}{RETURN}
19              {LET BUFFERA,@INT(@DATEVALUE(BUFFERA))}
20
21              /*CONVERT STRINGS TO INDEX KEY EXPRESSION SPECIFICATION*/
22  PAD         {DEFINE DB_FILE,FLD_NM,BUFFERA:VALUE}
23              {LET BUFFERA,@NDXPAD(DB_FILE,FLD_NM,BUFFERA)}
24
```

Figure 8-1

```
{ONERROR branch_location, message_location}
```

The branch_location portion of this command is the name of the macro to branch to if an error occurs. The message_location is the cell to which the {ONERROR} command should transfer the error message. The message_location argument is optional.

{USR_INP} begins by using {ONERROR} to set a branch to {FIN_RGT} whenever [CONTROL] and [BREAK] are pressed. The program line in cell FH6 sets up an infinite loop to obtain user input with the {?} command. If you recall from Chapter 6, the {?} command halts macro execution, thus allowing the user to type any combination of keystrokes, function keys, or arrow keys (to move the cell-pointer). The macro will continue after the user presses [RETURN]. This pause in macro execution allows the user to input text or numbers directly into cells. Pressing [RETURN] has no effect other than to end the {?} command. The tilde (~) in the cell after the {?} permanently enters user input into the spreadsheet.

The {BRANCH RGT_LP} in the cell after {?} sets up an infinite loop back to {?}. This looping allows as much editing and entering of data as needed. When [CONTROL] and [BREAK] are pressed, the macro breaks out of the infinite loop and branches to {FIN_RGT}.

After freezing the display and forcing a recalculation, the macro encounters a {DISPATCH} command. The {DISPATCH} command is necessary because a subroutine loses its ability to return to its calling function when [CONTROL] and [BREAK] are pressed. A {BRANCH} command could conceivably be used to return to the calling program, but a {DISPATCH} command allows the subroutine to be more generic.

If you recall from Chapter 6, the {DISPATCH macro_spec} command is the Lotus equivalent of a pointer to a function (subroutine) in C. When encountered in a macro program, the {DISPATCH} command causes a {BRANCH} to the macro that begins at the macro's name specified in the macro_spec cell, and the new macro begins executing. For example, suppose the SUB_PTR cell had the label "RGT_EXP". When the {DISPATCH SUB_PTR} command was encountered, {DISPATCH} would branch to the macro named "RGT_EXP" and begin executing.

To formalize the consequences of {DISPATCH} on the use of {USR_INP}, you must use a {LET} command in your macro programs to specifically change any pre-existing label in the SUB_PTR cell to the name of the macro you want your program to branch to prior to calling {USR_INP}. The example at the beginning of this section illustrates this detail.

{MK_STRNG}

FUNCTION: {MK_STRNG} checks whether the contents of BUFFERA cell is a string. If not, {MK_STRNG} converts the contents of the cell.

INPUT: Number, cell address, or formula containing the value to be converted.

OUTPUT: Original or converted string is placed in the cell called BUFFERA.

EXAMPLE: `{MK_STRNG BUFFERA}`

NOTES ON USAGE: Most operations that transfer data to databases have strict rules on whether values used are strings or numbers. Input for lot numbers, barcode numbers, etc., tend to be user-dependent. Some users may use numbers instead of strings. {MK_STRNG} is necessary to ensure strings where appropriate.

PROGRAM DESCRIPTION: {MK_STRNG} uses an @STRING @function to attempt to convert BUFFERA to a string. If BUFFERA is already a string, an ERR will be returned. The @STRING @function is nested within @ISERR. @ISERR will translate an ERR returned from @STRING into a "true" (that is, a one). If an ERR is not returned by @STRING, then @ISERR will return a "false" (zero). Thus, the value returned from @STRING is translated into something that an {IF} command can understand.

If @ISERR returns a true, the {IF} command proceeds to the {RETURN} and the subroutine returns to the calling function. No changes are made to the BUFFERA cell because the cell contents are already a string.

If @ISERR returns a false, the program proceeds to the next lower cell, converts BUFFERA to a string, places the string back into BUFFERA, and the subroutine returns to the calling program.

Alternatively, an @ISSTRING @function can be used in place of the @ISERR-@STRING combination.

To retrieve the value from BUFFERA, your calling program should contain a {LET} or {PUT} command to transfer the value of BUFFERA to the appropriate cell in your spreadsheet.

{MK_DT}

FUNCTION: {MK_DT} checks whether the contents of BUFFERA cell are a five-digit number conforming to the Lotus serial date format. If not, {MK_DT} converts the contents of the cell.

INPUT: Number, cell address, or formula containing the value to be converted.

OUTPUT: Original or converted four-digit number conforming to Lotus serial date format is returned to BUFFERA cell.

EXAMPLE: `{MK_DT @INDEX(RG_FM,3,11)}`

NOTES ON USAGE: Most operations involving databases have strict rules on whether date values used are strings or numbers. Templates are re-usable. If a user places a date string into a template, it must first be converted to its Lotus serial date equivalent. If a program has placed the date into the template, it may already be in the correct format. The {MK_DT} subroutine tests to see whether the date is already a Lotus date number. If not, the program converts it.

PROGRAM DESCRIPTION: {MK_DT} is similar in form and function to {MK_STRING}. {MK_DT} uses an @DATEVALUE @function to attempt to convert BUFFERA to a Lotus serial date number. If BUFFERA is already a number, an ERR will be returned. The @DATEVALUE @function is nested within @ISERR. @ISERR will translate an ERR returned from @DATEVALUE into a "true" (that is, a one). If an ERR is not returned by @DATEVALUE, then @ISERR will return a "false" (zero).

If @ISERR returns a true, the {IF} command proceeds to the {RETURN} command and the subroutine returns to the calling function. No changes are made to the BUFFERA cell because the cell already contains a serial date number.

If @ISERR returns a false, the program proceeds to the next lower cell, converts BUFFERA to a serial date number, places the number into BUFFERA, and the subroutine returns to the calling program.

The @INT function in MK_DT is very important. The @INT function will prevent potential rounding of times past noon if downstream utilities manipulate the date.

To retrieve the value from BUFFERA, your calling program should contain a {LET} or {PUT} command to transfer the value of BUFFERA to the appropriate cell in your spreadsheet.

{PAD}

FUNCTION: {PAD} constructs a character string from a specified value. Conversion is based on the field width in the database definition and is padded with spaces so that the resulting string matches the field width.

INPUT: DB_FILE: Character string for database file alias.
FLD_NM: Character string for field name.
BUFFERA: Number, cell address, or formula containing value to be padded.

OUTPUT: Character expression conforming to index key expression definition is returned to BUFFERA cell.

EXAMPLE:
```
{PAD "BATCH","ASSAY_TYPE",@INDEX(BCH_FM,2,5)}
{LET KEY,BUFFERA}
{PAD "BATCH","RUN_DT",@INDEX(BCH_FM,2,6)}
{LET KEY,KEY&BUFFERA}
{PAD "BATCH","BATCH",@INDEX(BCH_FM,6,5)}
{LET KEY,KEY&BUFFERA}

KEY          'DNA       33204  3
```

NOTES ON USAGE: When an index key expression involving multiple fields or date fields is used, conversions of numbers to strings and concatenation of the strings are performed by @BASE before the key value is placed into the index file. A concatenation process "adds" character strings together. When a search of the index is performed, @BASE compares a search string for the record to be found to the key values in the index file. When an exact match is found, the record number is returned. However, there is a very important part of the index key expression concatenation process that you need to be aware of. When concatenation is performed on strings shorter than the field length specified in the database definition, @BASE adds spaces to the string to extend the length of the string to conform to the field definition. Because an exact character-for-character match is required, reliable searches would not occur unless the appropriate number of spaces were reliably added. This process is called ***padding***. The {PAD} subroutine automates the padding process.

An important consequence of padding is: If numbers are used in index key expressions involving multiple fields, use the {PAD} subroutine to convert to strings and pad the appropriate number of spaces. ***DO NOT*** use @STRING to convert numbers to strings and then assemble keys containing the strings. However, if a number is used for a ***single*** field index key expression, the number ***cannot*** be a string.

An example of the importance of using {PAD} is shown at the beginning of this section. In Chapter 5, an index key expression containing a trio of index fields was created. The trio contained ASSAY_TYPE, RUN_DT, and BATCH. If you recall from Chapter 2, ASSAY_TYPE was the width equivalent of 10 characters, RUN_DT was 5, and BATCH was 3. If a record were being sought for an ASSAY_TYPE, RUN_DT, and BATCH of "DNA", 33204, and 3, respectively, merely concatenating the string would yield:

```
'DNA332043
```

This string would lead to "No Records Found". However, using {PAD} as illustrated in the example at the beginning of this section would yield the following string and would lead to a valid search:

```
'DNA        33204  3
```

Another advantage of using {PAD} is that it makes programs independent of data type. If a change in data type were made to a field definition, you would not need to change your macro program.

PROGRAM DESCRIPTION: {PAD} is based on the @BASE @NDXPAD @function. The syntax of @NDXPAD is:

```
@NDXPAD(DATABASE ALIAS,FIELD NAME,FIELD VALUE)
```

@NDXPAD constructs a character expression from a specified field value. The field value will be adjusted to conform to the format expected by the primary index. The intent of the function is to aid in the construction of index expressions involving multiple fields. The operation performed depends on the field type and current global settings as noted below:

- If the field type is Character and Case is set to Use-case, the value returned is padded on the right with spaces.
- If the field type is Character and Case is set to Ignore-case, the value returned is padded on the right with spaces and converted to upper-case.
- If the field type is Numeric, the value returned is converted to a right-justified character string and padded on the left with spaces.
- If the field type is Date, the value returned is a character string in the form YYYYMMDD with month and day space-filled to the left.

To retrieve the value from BUFFERA, your calling program should contain a {LET} or {PUT} command to transfer the value of BUFFERA to the appropriate cell in your spreadsheet.

DATABASE MENUS

Figure 8-2 shows two generic menus that are used in conjunction with exporting and importing data, respectively. They allow a user to specify which action to implement, based on the current status of the system.

OK_MU

FUNCTION: If a record already exists, the OK_MU menu can be used to ask a user whether to update or cancel overwriting the old record.

INPUT: None.

OUTPUT: None.

```
------ HY --------------------- HZ ------------------------------------ IA ----------------------------- IB -------------- IC --------------------------
 1
 2           /*RECORD EXISTS: ALLOW USER TO DECIDE*/
 3           /*WHETHER TO OVERWRITE EXISTING DATA*/
 4 OK_MU     CANCEL                   PROCEED
 5           RECORD EXISTS: CANCEL    RECORD EXISTS: PROCEED
 6           {BRANCH STOP}            {RETURN}
 7
 8           /*MENU TO GET NEXT OR PREVIOUS RECORD*/
 9 NPD_MU    NEXT                 PREVIOUS             DIFFERENT            QUIT
10           GET NEXT RECORD      GET PREVIOUS RECRDANOTHER RECORD          RETURN
11           {BRANCH NEXT_SRD}    {BRANCH PREV_SRD}    {DISPATCH SUB_PTR}{MENUBRANCH FN_MU}
12
```

Figure 8-2

EXAMPLE: `{IF RECRD<>0}{MENUCALL OK_MU}`

NOTES ON USAGE: If a record exists and a macro program is set to overwrite the record with new data, the OK_MU can be used to allow the user to specify whether to proceed with exporting the data. The OK_MU is a very important tool to alert a user to the potential destruction of existing data and allow the user to abort the process.

PROGRAM DESCRIPTION: OK_MU is a straightforward menu and is constructed according to the instructions provided in Chapter 7. If Cancel is chosen, OK_MU branches to {STOP}. If Proceed is chosen, OK_MU returns to the calling function. For this reason, a {MENUCALL} must be used in the calling function.

NPD_MU

FUNCTION: If more than one record in a database matches the index key, the NPD_MU menu can be used to allow a user to step back and forth through the matching records.

INPUT: NPD_MU requires a subroutine name in a cell named SUB_PTR.

OUTPUT: None.

EXAMPLE:
```
{LET SUB_PTR,"DO_SRD"}
{MENUBRANCH NPD_MU}
```

NOTES ON USAGE: NPD_MU is used by programs that look up a series of records with identical key values. When the records are found, the menu is used to allow the user to specify whether to get the next or previous record in the series or to move on to search for a different record. For example, if a program were retrieving several different assays on the same sample BARCODE number, NPD_MU could be used to obtain detailed information on each assay.

PROGRAM DESCRIPTION: NPD_MU is a straightforward menu and is constructed according to the instructions provided in Chapter 7. If Different is chosen, a {DISPATCH} command is executed. A {BRANCH} command could conceivably be used to return to the calling program, but a {DISPATCH macro_spec} command allows the menu to be more generic. When encountered in the menu, the {DISPATCH} command causes a {BRANCH} to the macro that begins at the macro's name specified in the macro_spec cell, and the new macro begins executing. For example, suppose the SUB_PTR cell had the label "DO_SRD". When the {DISPATCH SUB_PTR} command in NPD_MU was encountered, {DISPATCH} would branch to the macro named "DO_SRD" and begin executing.

To formalize the consequences of {DISPATCH} on the use of NPD_MU, you must use a {LET} command in your macro programs to specifically change any pre-existing label in the SUB_PTR cell to the name of the macro you want the NPD_MU menu to branch to prior to calling NPD_MU. The example at the beginning of this section illustrates this detail.

DATABASE AND INDEX FILE MANAGEMENT SUBROUTINES

Figure 8-3 shows subroutines that are involved in basic @BASE and database file functions. The following subsections describe the function, use, and programming features of each subroutine.

{CALL_BASE} and {RET_BASE}

FUNCTION: {CALL_BASE} calls up the @BASE menu; {RET_BASE} posts database changes and returns to the spreadsheet.

INPUT: None.

OUTPUT: None.

EXAMPLE:

```
{CALL_BASE}
    :
{RET_BASE}
```

NOTES ON USAGE: If you recall from Chapter 7, {CALL_BASE} and {RET_BASE} are subroutines that are customized by auto-executing macros and contain commands that are specific to the Release of Lotus software being used. This customization provides portability to the program for all current releases of @BASE and Lotus. The following are the commands that would appear for various Releases of Lotus 1-2-3 and Symphony:

- {CALL_BASE}
  - Lotus 1-2-3: {APP3}
  - Symphony 1.2 and 2.0: /@
  - Symphony 2.2: {TYPE}B/

```
--------FG--------FH-------FI-------FJ-------FK-------FL-------FM-------FN---
26
27            /*GENERIC SUBROUTINE UTILITIES (CONTINUED)*/
28
29  CALL_BASE /@                  /*CALL UP @BASE MENU*/
30
31  RET_BASE  FP{ESC}             /*POST INPUT & RETURN FROM @BASE*/
32
33            /*SELECT A PRIMARY INDEX FILE FOR DATABASE*/
34  SEL_NDX   {DEFINE DB_FILE,NDX}
35            {IF @NDXSELECT(DB_FILE,NDX)=0}{BRANCH ABORT}
36
37            /*RETURN DISPLAY TO NORMAL MODE AND RETURN TO MENU*/
38  STOP      {WINDOWSON}{PANELON}{CALC}
39            {MENUBRANCH FN_MU}
40
41            /*ERROR HAS OCCURRED; SIGNAL AND THEN STOP*/
42  ABORT     {BEEP}{BRANCH STOP}
43
44            /*SEE OF RECORD EXISTS; IF NOT CREATE A NEW RECORD*/
45            /*IMPORTANT NOTE: SEE TEXT REGARDING @BASE*/
46                      /*VERSIONS AND @DBAPP!!*/
47            /*NOTE: USE THIS SUBROUTINE FOR SINGLE KEY DBF'S ONLY*/
48  CHK_CRT   {DEFINE DB_FILE,FLD_NM,BUFFERA:VALUE}
49            {LET RECRD,@NDXFIRST(DB_FILE,BUFFERA)}
50            {IF RECRD<>0}{RETURN}
51            {LET RECRD,@DBAPP(DB_FILE)}
52            {IF @EXACT(RECRD,"@DBAPP(DB_FILE)")}{LET RECRD,0}
53            {LET TRAP,@DBUPD(DB_FILE,FLD_NM,RECRD,BUFFERA)}
54            {LET RECRD,@NDXFIRST(DB_FILE,BUFFERA)}
55
56            /*SEE IF RECORD EXISTS; IF YES, RETURN RECORD NUMBER*/
57  GET_REC   {DEFINE DB_FILE,BUFFERA:VALUE}
58            {LET RECRD,@NDXFIRST(DB_FILE,BUFFERA)}
```

Figure 8-3

- {RET_BASE}
  - Lotus 1-2-3: {ESC}
  - Symphony 1.2 and 2.0: FP{ESC}
  - Symphony 2.2: FP{ESC}{TYPE}S

PROGRAM DESCRIPTION: The {APP3} command in the Lotus 1-2-3 version of {CALL_BASE} is an @BASE macro command that automatically displays the @BASE menu. {APP3} is equivalent to pressing [ALT]-F9.

The {TYPE} command in {CALL_BASE} for Symphony 2.2 displays the Symphony Window Type menu so that the environment can be changed. The @BASE version for Symphony 2.2 requires an environment change to "BASE" before @BASE can be invoked. The {TYPE} command in {RET_BASE} displays the menu of window types again for a return to the Sheet environment.

The "FP" commands in the {RET_BASE} subroutines are critical. FP is an @BASE menu command that performs a *F*ile *P*ost. Without this command, you would lose any changes that have been made to database files as soon as you left the @BASE menu system. This peculiarity is due to a difference in the way that @BASE is controlled in manual versus macro mode. In manual mode, changes are automatically posted when you leave the menu system. When the @BASE menu is controlled by a macro, changes are not posted unless you specifically implement the post.

Also, because @BASE uses buffers to do reads and writes to disk, there may be data in buffers that have not been written to disk. If power had been turned off, it would result in missing data. *F*ile *P*ost is similar to closing and reopening a database file, except that it retains any current criteria.

Currently, the @BASE version for Lotus 1-2-3 does not have a File Post in the menu system, nor does it appear that one is needed. However, check future @BASE releases for Lotus 1-2-3 to see if a File Post has been added. If one has been provided, you should change the '\0 in Figure 7-1 to read:

```
{LET RET_BASE,"FP{ESC}"}
```

{SEL_NDX}

FUNCTION:	{SEL_NDX} selects the primary index file for an open database.	
INPUT:	DB_FILE:	Character string for database file alias.
	NDX:	Character string for index file alias to make primary.
OUTPUT:	None.	
EXAMPLE:	`{SEL_NDX "RESULTS","RES_BC"}`	

NOTES ON USAGE: If there are several index files open for a database file, you must specify which index file is primary. After a primary index file is selected, searches of the database will use the index key expression and key values in the primary index file. If you use the wrong index file as primary, searches will be ineffectual. For example, in Chapter 5 there were two index files created for the RESULTS.DBF database (RES_BC.NDX and RES_TRIO.NDX). If you had tried to search the index file by barcodes and RES_TRIO was primary, no matches would have been found.

PROGRAM DESCRIPTION: {SEL_NDX} is based on the @BASE @NDXSELECT @function. The syntax of @NDXSELECT is:

```
@NDXSELECT(DATABASE ALIAS,INDEX ALIAS)
```

@NDXSELECT selects the primary index file for an open database just like the *I*ndex *S*elect command on the @BASE Options menu. @NDXSELECT returns a 1 if the specified index file was open and was made primary, otherwise @NDXSELECT returns a 0. The {SEL_NDX} subroutine tests the returned value and branches to ABORT if a

0 is returned.

{STOP}

FUNCTION: {STOP} un-freezes the display, recalculates the spreadsheet, and branches program control to the function menu (FN_MU).
INPUT: None.
OUTPUT: None.
EXAMPLE: `{BRANCH STOP}`

NOTES ON USAGE: After each macro program is completed, {STOP} resets the system for the next operation.

PROGRAM DESCRIPTION: The {WINDOWSON}{PANELON} commands un-freeze the display. The {CALC} command forces a recalculation of the spreadsheet to ensure all cells display their current values. Thereafter, a {MENUBRANCH} command is used to branch back to the function menu. Without {MENUBRANCH}, the program would terminate.

{ABORT}

FUNCTION: {ABORT} signals that an error has occurred and then branches to {STOP}.
INPUT: None.
OUTPUT: None.
EXAMPLE: `{IF @ISERR(RECRD)#OR#RECRD=0}{BRANCH ABORT}`

NOTES ON USAGE: Throughout a program, {IF} tests are used to ensure that the system is configured correctly, records exist, etc. The {ABORT} subroutine is an error-handling routine that can be used to reset the system.

PROGRAM DESCRIPTION: {ABORT} is an adjunct to index and database file functions. {ABORT} is a simple subroutine that uses the {BEEP} command to signal that a simple error has occurred. {ABORT} then branches to {STOP} to reset the system and return to the function menu.

An alternative strategy would be to beep, display an error message in the indicator panel, pause the program, reset the system, and return to FN_MU. The following program can be used to implement this alternate system:

```
ABORT        {BEEP}{BEEP 2}{BEEP 3}{BEEP 2}{BEEP}
             {WINDOWSON}{PANELON}
             {INDICATE "ERROR"}
             {WAIT @NOW+@TIME(0,0,3)}
             {BEEP}
             {INDICATE}
             {BRANCH STOP}
```

To explain some of the features of the above program:

- The {BEEP} command sounds the personal computer's speaker. The numbers within the {BEEP} commands set the pitch of the speaker's sound.
- The {INDICATE} command is a convenient way to provide status reports and short prompts. The {INDICATE} command places a short string of characters into the mode indicator box. The mode indicator box is the small square box at the top right corner of the display. The syntax of this command is

  ```
  {INDICATE string}
  ```

  The string specification is optional. If used, it can be up to five characters long. If omitted, the {INDICATE} command will free the indicator box so that Lotus 1-2-3 or Symphony can regain command of it. When you use the {INDICATE} command with a string, the string remains on the screen, regardless of any subsequent changes that Lotus would normally make. It is important that the control panel be "on" before the mode indicator is updated. Therefore, if you have issued a {PANELOFF} command to freeze the panel, you must issue a {PANELON} command to update the mode indicator.
- The {WAIT} command suspends all processing until a certain time has been reached. It then continues with the next program step. In the above program, the {WAIT} command begins by getting the current time using the @NOW @function. It then evaluates the @TIME @function and adds its value to the value of @NOW. This new value is a serial number for a future timepoint. The syntax of @TIME is

  ```
  @TIME(hours, minutes, seconds)
  ```

  The @TIME function translates the three arguments into a decimal fraction equivalent to the time of day. This decimal fraction is between 0.0000000000 (midnight) and 0.9999884259 (11:59:59 p.m.). The @TIME @function in the {WAIT} command translates a delay time from seconds (or minutes) into a form that can be added to the @NOW's. For example, if you were to add 5 (for 5 seconds) directly to the value from @NOW, you would be adding 5 days! The @TIME @function, therefore, is essential because it translates the number of seconds into decimal fractions that the program can add to @NOW's time. Once the serial number is determined by addition, the {WAIT} command suspends all processing until the serial number of the personal computer's internal clock reaches the calculated serial number. All processing is suspended until the time indicated by the serial number argument has been reached. It is important to realize that the {WAIT} command is performing a "wait ***until*** a particular time" and not a "wait ***for*** a certain amount of time".

{CHK_CRT}

FUNCTION: {CHK_CRT} checks to see if a record exists for a given key value. If it does, {CHK_CRT} returns the record number. If not, {CHK_CRT} creates a new record and returns the new record number.

INPUT:
- DB_FILE: Character string for database file alias.
- FLD_NM: Character string for field name used in an index. If a new record is created, a key value for the record will be placed into the field specified by FLD_NM.
- BUFFERA: Number, cell address, or formula containing the value to be used as a key when searching the index file.

OUTPUT: The record number is placed into RECRD cell.

EXAMPLE: `{CHK_CRT "SAMPLES","BARCODE",BCODE}`

NOTES ON USAGE: Before a macro program can place data into a record, the record must first exist and the record number must be made available to the macro program. {CHK_CRT} checks to see if a record exists for a given key value. If a record does not exist, {CHK_CRT} creates a new record and places the value of the key into the FLD_NM (Field Name) field. The record number of the existing or new record is placed in the RECRD cell of the spreadsheet.

{CHK_CRT} can only be used for single key indexes. If an index has multiple keys, you should use the {GET_REC} with a properly padded key to check whether a record exists and either the {DB_EXP} or {KEY_EXP} subroutine to create a new record if a record is not found.

Prior to using {CHK_CRT}, your calling program must set the primary index file with {SEL_NDX}.

PROGRAM DESCRIPTION: The check for existence of a record is based on the @BASE @NDXFIRST @function. The syntax of @NDXFIRST is:

```
@NDXFIRST(DATABASE ALIAS,SEARCH VALUE)
```

@NDXFIRST searches the primary index file for the first index key value that matches the search key value. This command is also known as an index seek. The function of @NDXFIRST is similar to using @DBFIRST, but is much faster because it uses the primary index file to perform the lookup. The format of the search key value must be the same as that of the index key expression. Spaces within the index key expression must be included. For this reason, when working with strings or index key equations with multiple field names, {PAD} must be used before using @NDXFIRST. If a record is located, the record number is returned. If a record that matches the search key value is not found, a zero is returned.

The {IF} command in cell FH50 tests whether a record number (i.e., a non-zero) is returned. If a record number is returned, the record exists and {CHK_CRT} returns to the calling function. The record number can be obtained from the RECRD cell in the spreadsheet.

If a zero is returned, {CHK_CRT} creates a new record. This method illustrates a very important portability feature that you will need to implement. Some versions of @BASE use an @DBAPP @function to add a blank record to the end of a database file. The function returns the record number of the newly appended record. Once the new record has been appended with @DBAPP, the fields in the record can be assigned values using @DBUPD.

Unfortunately, not all versions of @BASE use this technique. With some versions, a record is created and appended by using a record number of zero in an @DBUPD function. @DBAPP is not available with these versions.

The portability system is based on a peculiarity in the functionality of the {LET} commands. If a {LET} command contains a cell designation or expression not recognized by Lotus, the {LET} command will treat the cell designation or expression as a string and place the string in the target cell. If @DBAPP exists, the record number will be placed in RECRD. However, if @DBAPP is not a viable @function, it will not be recognized and the {LET} command in cell FA57 will place the string "@DBAPP(DB_FILE)" into RECRD. By testing for the presence of the string in RECRD using an @EXACT function, you can determine whether @DBAPP exists in the version of @BASE. If non-existent, a zero is placed into the RECRD cell for later use by the @DBUPD in cell FH53.

For versions of @BASE that do not have @DBAPP, the @DBUPD @function will create a new record if the record number is zero. @DBUPD has the following syntax:

```
@DBUPD(DATABASE ALIAS,FIELD NAME,RECNO,VALUE)
```

@DBUPD replaces the contents of the field FIELD NAME in the record RECNO with the field value VALUE. As configured in {CHK_CRT}, the field name is the one used in the index file. Therefore, the new record automatically contains the key value.

Because @DBUPD does not return the record number, the macro ends by determining the record number using @NDXFIRST and placing the record number into RECRD. To retrieve the value of RECRD, your calling program should contain a {LET} or {PUT} command to transfer the value of RECRD to the appropriate cell in your spreadsheet.

{GET_REC}

FUNCTION: {GET_REC} determines whether a particular record exists for a given key value. If it does, {GET_REC} returns the record number.

INPUT:
- DB_FILE: Character string for database file alias.
- BUFFERA: Number, cell address, or formula containing value to be used as a key value when searching the index file.

OUTPUT: Record number placed into RECRD cell.

EXAMPLE: `{GET_REC "SAMPLES",BUFFERA}`

NOTES ON USAGE: Before a record can be updated or the contents of the record's fields can be retrieved, the record number must be made available. Similarly, the first occurrence of a record must be identified before @NDXNEXT and @NDXPREV @functions can be used. {GET_REC} supports these requirements and places the record number into the RECRD cell.

PROGRAM DESCRIPTION: {GET_REC} is based on @NDXFIRST. For a description of @NDXFIRST, see the previous section on {CHK_CRT}. As with {CHK_CRT}, the primary index file must be set with {SEL_NDX} prior to using {GET_REC}.

DATABASE RECORD MANIPULATION SUBROUTINES

Figure 8-4 shows subroutines that manage the transfer of data to and from databases. The following subsections describe the function, use, and programming features of each subroutine.

{UPDATE}

FUNCTION: {UPDATE} modifies a record in a database by changing the entry in one of the record's fields.

INPUT:
- DB_FILE: Character string for database file alias.
- FLD_NM: Cell address or formula containing name of field to be modified.
- BUFFERA: Number, cell address, or formula containing value to be placed in a field being modified.

OUTPUT: If successful, a 1 is placed into the TRAP cell. If unsuccessful, a 0 or ERR is placed into the TRAP cell. A zero is returned if

- an attempt is made to fill a field with an incorrect data type.
- an invalid database alias or field name is supplied.
- the record number is invalid.

EXAMPLE: `{UPDATE "SAMPLES",BUFFERD,@INDEX(SID_FM,BUFFERB,BUFFERC)}`

NOTES ON USAGE: {UPDATE} is one (of several) subroutines that can be used to update fields. The primary purpose of {UPDATE} is for situations where a template of data is not arranged as a table comprised of contiguous columns of data containing a row of field name labels above the data. Typically, {UPDATE} is placed in a {FOR} loop and fields are changed using a table of field names and an @INDEX @function. Data are retrieved from the spreadsheet with @INDEX @functions and given to {UPDATE} for export to the database. {UPDATE} offers great flexibility in data arrangement, but is slower than using {DB_EXP} or {KEY_EXP}.

PROGRAM DESCRIPTION: {UPDATE} is based on the @BASE @DBUPD @function.

```
--------FG--------FH-------FI-------FJ-------FK-------FL-------FM-------FN---
61
62              /*GENERIC SUBROUTINE UTILITIES (CONTINUED)*/
63
64              /*PLACE DATA INTO DATABASE, UPDATING RECORD*/
65  UPDATE      {DEFINE DB_FILE,FLD_NM:VALUE,BUFFERA:VALUE}
66              {LET TRAP,@DBUPD(DB_FILE,FLD_NM,RECRD,BUFFERA)}
67
68              /*GET DATA FROM ONE OF THE RECORD'S FIELDS*/
69                      /*PLACE VALUE INTO BUFFERA*/
70  GET_FLD     {DEFINE DB_FILE,FLD_NM:VALUE,RECRD:VALUE}
71              {LET BUFFERA,@DBFLD(DB_FILE,FLD_NM,RECRD)}
72
73              /*GET NEXT OCCURRENCE OF RECORD*/
74  GET_NEXT    {DEFINE DB_FILE,BUFFERA:VALUE}
75              {LET RECRD,@NDXNEXT(DB_FILE,BUFFERA)}
76
77              /*GET PREVIOUS OCCURRENCE OF RECORD*/
78  GET_PREV    {DEFINE DB_FILE,BUFFERA:VALUE}
79              {LET RECRD,@NDXPREV(DB_FILE,BUFFERA)}
80
81              /*IMPORT A TABLE FROM A DATABASE*/
82              /*USING FIELD NAMES IN TEMPLATE*/
83  KEY_IMP     {DEFINE IMP_FILE,IMP_LBLS:VALUE,IMP_KEY:VALUE}
84              {CALL_BASE}OIDK   /*CALL UP @BASE; OPTIONS,INDEX,DATA,KEY-IMPORT*/
85  IMP_FILE    RESULTS           /*DATABASE FILE NAME TO IMPORT DATA FROM*/
86              ~L                /*LABELS*/
87  IMP_LBLS    CERT_HDR          /*RANGE CONTAINING IMPORT LABELS*/
88              ~                 /*CARRIAGE RETURN*/
89  IMP_KEY     101               /*IMPORT KEY TO USE (E.G., BARCODE NMBR)*/
90              ~E{RET_BASE}      /*EXECUTE; RETURN FROM @BASE*/
91
92              /*EXPORT TABLE OF DATA FROM A TEMPLATE*/
93              /*TO A DATABASE; RECNO PERMITTED AS A FIELD NAME*/
94  DB_EXP      {DEFINE EXP_FILE,EXP_RNG:VALUE}
95              {CALL_BASE}DTE    /*CALL UP @BASE; DATA,TRANSFER,EXPORT*/
96  EXP_FILE    RESULTS           /*DATABASE FILE NAME TO EXPORT DATA TO*/
97              ~                 /*CARRIAGE RETURN*/
98  EXP_RNG     FL_REP            /*RANGE TO EXPORT*/
99              ~{RET_BASE}       /*CARRIAGE RETURN; RETURN FROM @BASE*/
100
101             /*EXPORT TABLE OF DATA FROM A TEMPLATE*/
102             /*TO A DATABASE USING OPTIONS INDEX*/
103             /*NOTE: FOR THIS FUNCTION, CANNOT HAVE Recno AS A FIELD!*/
104 KEY_EXP     {DEFINE EXP_FILE,EXP_RNG:VALUE}
105             {CALL_BASE}OIDE   /*CALL UP @BASE; OPTIONS,INDEX,DATA,EXPORT*/
106 KEY_FILE    RESULTS           /*DATABASE FILE NAME TO EXPORT DATA TO*/
107             ~                 /*CARRIAGE RETURN*/
108 KEY_RNG     FL_REP            /*RANGE TO EXPORT*/
109             ~UE{RET_BASE}     /*[RETURN];UPDATE,EXECUTE;RETURN FROM @BASE*/
```

Figure 8-4

The syntax of @DBUPD is:

```
@DBUPD(DATABASE ALIAS,FIELD NAME,RECNO,VALUE)
```

@DBUPD replaces the contents of the field FIELD NAME in the record RECNO with VALUE. Prior to using @DBUPD you must use {GET_REC} to determine the number of the record to be updated.

{GET_FLD}

FUNCTION: {GET_FLD} places the contents of a specified field of a database record into BUFFERA of the spreadsheet.

INPUT:
- DB_FILE: Character string for database file alias.
- FLD_NM: Cell address or formula containing name of field to be retrieved.
- RECRD: Number, cell address, or formula containing record number.

OUTPUT: Contents of field placed into BUFFERA cell.

EXAMPLE: `{GET_FLD "SAMPLES",@INDEX(SI_TAB,0,INC),RECRD}`

NOTES ON USAGE: {GET_FLD} is one (of several) subroutines that can be used to retrieve data from fields in a database. The primary purpose of {GET_FLD} is for situations where the data will not be placed in a table containing contiguous columns. Typically, {GET_FLD} is placed in a {FOR} loop and fields are changed using a table of field names and an @INDEX @function. Data are retrieved from a database and placed into the template with a {PUT} command. {GET_FLD} offers great flexibility in data arrangement, but is slower than using {KEY_IMP}.

PROGRAM DESCRIPTION: {GET_FLD} is based on the @BASE @DBFLD @function. The syntax of @DBFLD is:

```
@DBFLD(DATABASE ALIAS,FIELD NAME,RECNO)
```

@DBFLD returns the contents of the field FIELD NAME for the record RECNO. Prior to using @DBFLD you must use {GET_REC} to determine the number of the record to be retrieved. To retrieve the value from BUFFERA, your calling program should contain a {LET} or {PUT} command to transfer the value of BUFFERA to the appropriate cell in your spreadsheet.

{GET_NEXT}

FUNCTION: {GET_NEXT} starts at the current record, searches forward for the next record that matches the key value, and places the record number of the next occurrence into the BUFFERA cell.

INPUT: DB_FILE: Character string for database file alias.
BUFFERA: Number, cell address, or formula containing value to be used as a key value when searching the index file.
OUTPUT: Next record number placed into the cell called RECRD.
EXAMPLE: `{GET_NEXT "SAMPLES",BCODE}`

NOTES ON USAGE: {GET_NEXT} and {GET_PREV} are typically used for situations in which several results on the same sample are archived. For example, for a single sample barcode there may have been several different assays performed. {GET_NEXT} and {GET_PREV} can be used to move back and forth through all occurrences of different assay types, assay dates, etc, on the sample. Before {GET_NEXT} and {GET_PREV} can be used, you must first use the {GET_REC} subroutine to locate the first occurrence of the record matching the search key value.

PROGRAM DESCRIPTION: {GET_NEXT} is based on the @BASE @NDXNEXT @function. The syntax of @NDXNEXT is:

```
@NDXNEXT(DATABASE ALIAS,SEARCH VALUE)
```

@NDXNEXT searches the primary index file for the next record in a data file with an index key field that matches the supplied search key value. The function of @NDXNEXT is similar to using @DBNEXT, but it is much faster because it uses the primary index file to perform the lookup. The format of the search key value must be the same as that of the index key expression. Spaces within the index key expression must be included. For this reason, {PAD} must be used before using @NDXNEXT with strings or index key expressions with multiple field names. If a record is located, the record number is returned. If a record that matches the search key value is not found, a zero is returned.

After the record number is found, the record number is placed into the RECRD cell. The field contents can then be imported using {GET_FLD} or {KEY_IMP}.

{GET_PREV}

FUNCTION: {GET_PREV} starts at the current record, searches backward for the next record that matches the key value, and places the record number of the previous occurrence into the BUFFERA cell.
INPUT: DB_FILE: Character string for database file alias.
BUFFERA: Number, cell address, or formula containing value to be used as a key value when searching the index file.
OUTPUT: Previous record number placed into the cell called RECRD.
EXAMPLE: `{GET_PREV "RESULTS",BCODE}`

NOTES ON USAGE: {GET_NEXT} and {GET_PREV} are typically used for situations in which several results on the same sample are archived. For example, for a single sample barcode there may have been several different assays performed. {GET_NEXT} and

{GET_PREV} can be used to move back and forth through all occurrences of different assay types, assay dates, etc, on the sample. Before {GET_NEXT} and {GET_PREV} can be used, you must first use the {GET_REC} subroutine to locate the first occurrence of the record matching the search key value.

PROGRAM DESCRIPTION: {GET_PREV} is based on the @BASE @NDXPREV @function. The syntax of @NDXPREV is:

```
@NDXPREV(DATABASE ALIAS,SEARCH VALUE)
```

@NDXPREV searches the primary index file for the previous record in a data file with an index key field that matches the supplied search key value. The function of @NDXPREV is similar to using @DBPREV, but it is much faster because it uses the primary index file to perform the lookup. The format of the search key value must be the same as that of the index key expression. Spaces within the index key expression must be included. For this reason, {PAD} must be used before using @NDXPREV with strings or index key expressions with multiple field names. If a record is located, the record number is returned. If a record that matches the search key value is not found, a zero is returned.

After the record number is found, the record number is placed into the RECRD cell. The field contents can then be imported using {GET_FLD} or {KEY_IMP}.

{KEY_IMP}

FUNCTION: {KEY_IMP} transfers records from a database file to a template table based on a key value. Selected fields (specified by a row of field name labels in the template) are imported.

INPUT:
- IMP_FILE: Character string for database file alias.
- IMP_LBLS: Cell address or formula containing the name of the range bearing labels that specify the fields to import.
- IMP_KEY: Cell address or formula containing a value to be used as a key when searching the index file to determine which records are to be imported. Value ***MUST*** be a character string, ***NOT*** a number.

OUTPUT: All active records in the database file with the specified index key value are transferred to the template. If "Recno" is in the IMP_LBLS range, record numbers are also imported. Records are organized in rows and columns.

EXAMPLE: `{KEY_IMP "RESULTS",@UPPER("CERT_HDR"),@STRING(BCODE,0)}`

NOTES ON USAGE: The {KEY_IMP} subroutine is used for importing tables of data from database files. {KEY_IMP} is much faster than using {GET_FLD} subroutines to obtain individual data fields from databases and then transferring the data to a template using {PUT} commands. This speed difference is due to the relatively slow performance of

{PUT} commands and is especially noticeable when importing large amounts of data. However, {KEY_IMP} is more restrictive than {GET_FLD} because the data are placed in a table.

To use {KEY_IMP}, a row of field labels must be set up in the spreadsheet before the operation begins. The order in which you specify the fields in the spreadsheet is the order in which they will appear in the spreadsheet.

If Recno is the first label in the row, record numbers are also imported. You should import the record numbers into the spreadsheet if you plan to modify records and export the records back to the database file using {DB_EXP}. If you include record numbers when you export records to a database, the database records with matching numbers are updated by the spreadsheet records. If you do not include record numbers when exporting data, all records are appended to the end of the database instead of being used to update existing records.

{KEY_IMP} overwrites everything in the cells that fall within the output range. So, be sure to leave enough space below the row of import definition labels to account for the number of records being imported.

If you try to use {KEY_IMP} and receive an "Unknown Field Type" message, the message sometimes means that you have the wrong spelling for one of the field names in the spreadsheet. However, most often the message is received because field names in the specification range have all upper-case characters for one or more virtual fields. For virtual fields, you must use lower-case characters. If you are not able to retrieve records, it is often a sign that you have not used {SEL_NDX} to select the appropriate primary index file.

PROGRAM DESCRIPTION: {KEY_IMP} uses the @BASE ***O***ptions ***I***ndex ***D***ata ***K***ey-Import menu command sequence to import a table of data from a database to a template. ***O***ptions ***I***ndex ***D***ata ***K***ey-Import is much faster than ***D***ata ***T***ransfer ***I***mport ***L***abels because it uses the primary index file to provide faster access to the database records.

The programming in the {KEY_IMP} macro is relatively straightforward except for the IMP_FILE, IMP_LBLS, and IMP_KEY cells. The define statement contains names for these cells. When arguments are passed to {DB_EXP}, the values of the arguments are placed into the cells specified in the {DEFINE} statement. In this case, the cell names in {DEFINE} are IMP_FILE, IMP_LBLS, and IMP_KEY. Because the cells are in the path of macro execution, they become part of the macro. In these cases, the cells contain specifications for the name of the database file to import the data from, the name of the range containing the import labels, and the key value of the record(s) to import, respectively. After each of these cells, a "~" is used to transmit a [RETURN] to the @BASE menu system.

If an index file has an index key expression with multiple fields, @BASE prompts for an entry for each key value in the index key expression. For example, BCH_TRIO.NDX is an index file containing three keys. When using BCH_TRIO.NDX manually, @BASE will prompt for three values. {KEY_IMP} can be used for index key expressions containing multiple fields if {PAD} is used for each string and the strings are concatenated with "~" after all but the last string. The following is an example macro sequence and final concatenated string:

```
{PAD "BATCH","ASSAY_TYPE",@INDEX(BCH_FM,2,5)}
{LET BUFFERD,BUFFERA&"~"}
{PAD "BATCH","RUN_DT",@INDEX(BCH_FM,2,6)}
{LET BUFFERD,BUFFERD&BUFFERA&"~"}
{PAD "BATCH","BATCH",@INDEX(BCH_FM,6,5)}
{LET BUFFERD,BUFFERD&BUFFERA}

BUFFERD        'DNA       ~33204~  3
```

{DB_EXP}

FUNCTION: {DB_EXP} transfers records from a spreadsheet table to a database file. If an exported record number matches a record number in the database file, the exported record overwrites the existing record. If record numbers are not included in the table, or if the record numbers do not match any record numbers in the database file, the exported records are appended to the end of the file. Selected fields (specified by a row of field name labels in the template) are exported.

INPUT: EXP_FILE: Character string for database file alias.
EXP_RNG: Cell address or formula containing name of range bearing data to be exported.

OUTPUT: None.

EXAMPLE: `{DB_EXP "RESULTS",@UPPER("FL_REP")}`

NOTES ON USAGE: There are two subroutines for exporting data from a spreadsheet to a database file: {DB_EXP} and {KEY_EXP}. Both subroutines are much faster than obtaining data from a template using @INDEX functions and transferring the data to a database with {UPDATE}. However, {DB_EXP} and {KEY_EXP} are more restrictive than {UPDATE} because the data must be rigidly structured in a table before they can be used.

There are subtle, but important, differences in the way in which the {DB_EXP} and {KEY_EXP} subroutines export data. Both subroutines export data from a table in a spreadsheet. The table is formatted such that a row of labels corresponding to field names in the database definition is at the top of the table. The names of the fields correspond to the columns being exported. The major difference in the two subroutines arises from whether record numbers are used. The Recno field name is valid for {DB_EXP}, but not for {KEY_IMP}.

Thus, {DB_EXP} can be used to target specific records when multiple records have the same key value. If the record number was determined using {GET_REC} and was placed in a column headed by "Recno," {DB_EXP} would overwrite the existing data of the target record. Thus, if your calling program finds record numbers at the beginning of the program, the program can be used to overwrite existing data.

{DB_EXP} can also be used to create several records with the same key value. A common example would be one in which several different assay types were performed on the same sample. If the index key is BARCODE, {KEY_EXP} would overwrite the first occurrence, but {DB_EXP} would create new records if the Recno column was not present or if a zero was placed in the Recno column.

{KEY_EXP} is the best choice to use if you want to update (overwrite) existing records. {KEY_EXP} exports data based on key values in the table. To use {KEY_EXP}, the first entries in the field label row must contain the field names used in the current primary index key expression. (Again, you ***CANNOT*** use a Recno field for {KEY_EXP}).

Both {DB_EXP} and {KEY_EXP} are the best ways to create records for index files that have index key expressions containing multiple keys. To do so, place the key values in the first columns of the table and export the data. If using {DB_EXP}, either do not use a Recno field name or place zeros in the Recno column.

To use either program, a row of field labels must be set up in the spreadsheet before operation begins. Both subroutines also need a range to export. The range must start at the row containing the field names and extend down the data to be exported. The range cannot be open-ended. If the range is larger than the table of data to be exported, empty records will be created. If the range is smaller than the table of data, not all of the records will be exported.

PROGRAM DESCRIPTION: {DB_EXP} is based on the @BASE ***D***ata ***T***ransfer ***E***xport menu command sequence. When exporting data using ***D***ata ***T***ransfer ***E***xport, @BASE looks at the Recno field for each record in the spreadsheet and compares it against the database. If there is already a record with that number in the database, the record is replaced with the one from the spreadsheet. If the record number does not match, a new record is appended to the end of the database file using data from the spreadsheet. Therefore, {DB_EXP} can be used to both change existing records and add new records to a database in one operation.

A common use of ***D***ata ***T***ransfer ***E***xport is in conjunction with ***D***ata ***T***ransfer ***I***mport. ***D***ata ***T***ransfer ***I***mport is used to transfer records to the spreadsheet so that they can be modified and transferred back to the database file with ***D***ata ***T***ransfer ***E***xport. Similarly {DB_EXP} can be used in conjunction with {DB_IMP} to implement the same process.

The programming in the {DB_EXP} macro is relatively straightforward except for the EXP_FILE and EXP_RNG cells. The define statement contains names for these cells. When arguments are passed to {DB_EXP}, the values of the arguments are placed into corresponding cells. Because the cells are in the path of macro execution, they become part of the macro. In these cases, the cells contain specifications for the name of the database file to export the data to and the range to export, respectively. After each cell, a " ~ " is used to transmit a [RETURN] to the @BASE menu system.

The {RET_BASE} subroutine at the end of {DB_EXP} is used to force the updated information to be written to the database file. Without posting the new information, the new information may be lost.

{KEY_EXP}

FUNCTION: {KEY_EXP} transfers records from a spreadsheet table to a database file.

Selected fields (based on a row of field name labels in the template) are exported.

INPUT: KEY_FILE: Character string for database file alias.
KEY_RNG: Cell address or formula bearing the name of the range containing the data to be exported.

OUTPUT: None.

EXAMPLE: `{KEY_EXP "RESULTS",@UPPER("FL_REP")}`

NOTES ON USAGE: For a discussion of when and how to use {KEY_EXP}, see "Notes on Usage" under {DB_EXP}. {KEY_EXP} has two extra requirements:

- The first entries in the field labels row must contain the fields used in the current primary index key expression.
- Recno is not a valid entry in the field labels row.

Only the first record found that matches the index key fields will be updated. Prior to using {KEY_EXP}, your calling program must set the primary index file with {SEL_NDX}.

PROGRAM DESCRIPTION: {KEY_EXP} is similar in form and function to {DB_EXP}. {KEY_EXP} is based on the ***O**ptions **I**ndex **D**ata **E**xport* command sequence. ***O**ptions **I**ndex **D**ata **E**xport* is similar to the ***D**ata **T**ransfer **E**xport* command. However, the operation is much faster because it uses the primary index file to provide faster access to the database records.

DESIGNING YOUR OWN SUBROUTINE UTILITIES

There are two other subroutines that are often useful. The first subroutine transfers records from an open database file to a spreadsheet table based on key values in the spreadsheet. The operation is based on ***O**ptions **I**ndex **D**ata **I**mport*. A row of field labels must be set up in the spreadsheet before operation begins. The first entries in the field labels row must contain the fields used in the current primary index key expression. For each row containing key values, @BASE finds the first record that matches the key values and transfers its fields to the spreadsheet. Only fields that match the entries in the row of field labels are moved into the spreadsheet.

The second subroutine is useful for end-of-month reports. This subroutine is based on the @BASE @DBCNT @function. The syntax of @DBCNT is:

```
@DBCNT(DATABASE ALIAS,CRITERIA)
```

@DBCNT returns the number of records in DATABASE ALIAS that meet CRITERIA. For example, suppose you wanted to issue a report of the total number of assays completed in December 1990. You could use the following @function to obtain the count (assuming Lotus serial date numbers were used in the database file):

```
@DBCNT("RESULTS,"RUN_DT>33208#AND#RUN_DT<33238")
```

If you wanted to provide more detailed information on the month by including sample status, you could use the following @DBCNTs:

```
@DBCNT("RESULTS,"RUN_DT>33208#AND#RUN_DT<33238#AND#R_STATUS='COMP'")
@DBCNT("RESULTS,"RUN_DT>33208#AND#RUN_DT<33238#AND#R_STATUS='INC'")
```

@BASE provides a number of other statistical and database @functions that allow criteria to be used when creating composite reports. The following are some examples. For a more detailed description of these @functions, see Appendix D.

- @DBCNT(DATABASE ALIAS,CRITERIA)
- @DBMAX(DATABASE ALIAS,FIELD NAME,CRITERIA)
- @DBMIN(DATABASE ALIAS,FIELD NAME,CRITERIA)
- @DBSUM(DATABASE ALIAS,FIELD NAME,CRITERIA)
- @DBFIRST(DATABASE ALIAS,CRITERIA)
- @DBNEXT(DATABASE ALIAS,CRITERIA)
- @DBPREV(DATABASE ALIAS,CRITERIA)
- @NDXFIRST(DATABASE ALIAS,SEARCH VALUE,CRITERIA)
- @NDXNEXT(DATABASE ALIAS,SEARCH VALUE,CRITERIA)
- @NDXPREV(DATABASE ALIAS,SEARCH VALUE,CRITERIA)

It would be good practice to design and create these subroutines. Call the subroutines {IDI_IMP} and {DB_COUNT}, respectively. Use the {KEY_IMP}, {DB_EXP}, and {KEY_EXP} subroutines as models for construction of {IDI_IMP} and {GET_FLD} as a model for {DB_COUNT}.

For {DB_COUNT}, use a {DEFINE} command to specify the following:

```
{DEFINE DB_FILE,BUFFERB:VALUE}
```

By passing the criteria as a string placed in BUFFERB, the {DB_COUNT} subroutine can remain generic. That is, {DB_COUNT} can work with any field name or group of field names. A string concatenation can be used to create the string in BUFFERB. (See Chapter 9.)

The {LET} command in the {DB_COUNT} subroutine should be

```
{LET BUFFERA,@DBCNT(DB_FILE,BUFFERB)}
```

Using the above {LET} command, the number of records meeting the criteria will be placed into the BUFFERA cell. To retrieve the value from BUFFERA, your calling program should contain a {LET} or {PUT} command to transfer the value of BUFFERA to the appropriate cell in your spreadsheet.

Place the subroutines below {KEY_EXP} in column FH of the example spreadsheet.

CREATING THE EXAMPLE SUBROUTINE UTILITIES

Create the subroutines and menus in Figures 8-1 through 8-4 by entering the text into your spreadsheet. To activate the subroutines, issue the /***R***ange ***N***ame ***L***abel ***R***ight command and highlight the cells containing the subroutine names in Column FG. Press [RETURN]. Repeat for the cells containing the menu names in Column HY.

You will not be able to use the subroutines at this time because you will need both calling programs and a scratch-pad of cells. These components are the subject of the next chapter.

SUMMARY OF WHAT HAS BEEN ACCOMPLISHED

This chapter presented a library of subroutines that can be used as building blocks within macro programs tailored to specific LIM system tasks. The subroutines are generic, meaning that they can cater to a wide variety of calling programs. However, care must be maintained when using the subroutines.

- You must be careful when passing arguments to the subroutines. Arguments ***MUST*** be in the correct format. Most problems are due to using a string argument when a value is needed, and vice versa.
- You must be careful to check the spelling of each argument before using the subroutines.
- Data types must be strictly adhered to when used as key values and when being input into database fields.
- The primary index file needs to be selected prior to using index subroutines.
- {GET_REC} needs to be used before {UPDATE}, {GET_FLD}, {GET_NEXT}, {GET_PREV}, etc.
- You must be very careful when using numbers. Sometimes numbers need to be converted to strings and/or padded before use. At other times, they must be numbers.
- You must use {PAD} when using index key expressions that have multiple or date fields.
- You must be careful when creating field name rows. If field names are virtual, they must contain lower-case characters.

{GET_FLD} and {UPDATE} are slower than {KEY_IMP}, {DB_EXP} and {KEY_EXP}, but provide more flexibility in data placement.

WHAT'S NEXT?

The next chapter explains how to create programs that transfer data to and from database files. The programs provided in Chapter 9 form the basis for an operational LIM system and can be easily modified to most laboratory applications.

9

LIM System Programming

It was pointed out at the beginning of Chapter 8 that most programs that supply functionality to the menu system created in Chapter 7 could be separated into the following functional categories. They

- accept user input to obtain specifications for the record(s) to be imported or exported.
- accept user input to enter or edit data.
- format data to conform to the specifications used in database definitions.
- select the primary index file(s) to use.
- determine existing record numbers or create new records.
- retrieve data from a template and transfer them to a database if data are exported.
- retrieve record number-specific data from a database and transfer the data to a template if data are imported.

The goal of LIM system programs is to move data into and out of the templates created as a result of Chapter 3 as quickly and accurately as possible, while using the least amount of memory. This chapter provides programs to illustrate how to combine the subroutine building blocks created in Chapter 8 to create a system that comes as close to meeting these goals as possible.

FUNCTION DISTRIBUTION PROGRAMS

The goal of the menu system described in Chapter 7 and illustrated in Figure 9-1 is to allow users to proceed down the hierarchy of menus until the user "zeros" in on the task that needs to be performed; at which time an appropriate macro is called upon to do the

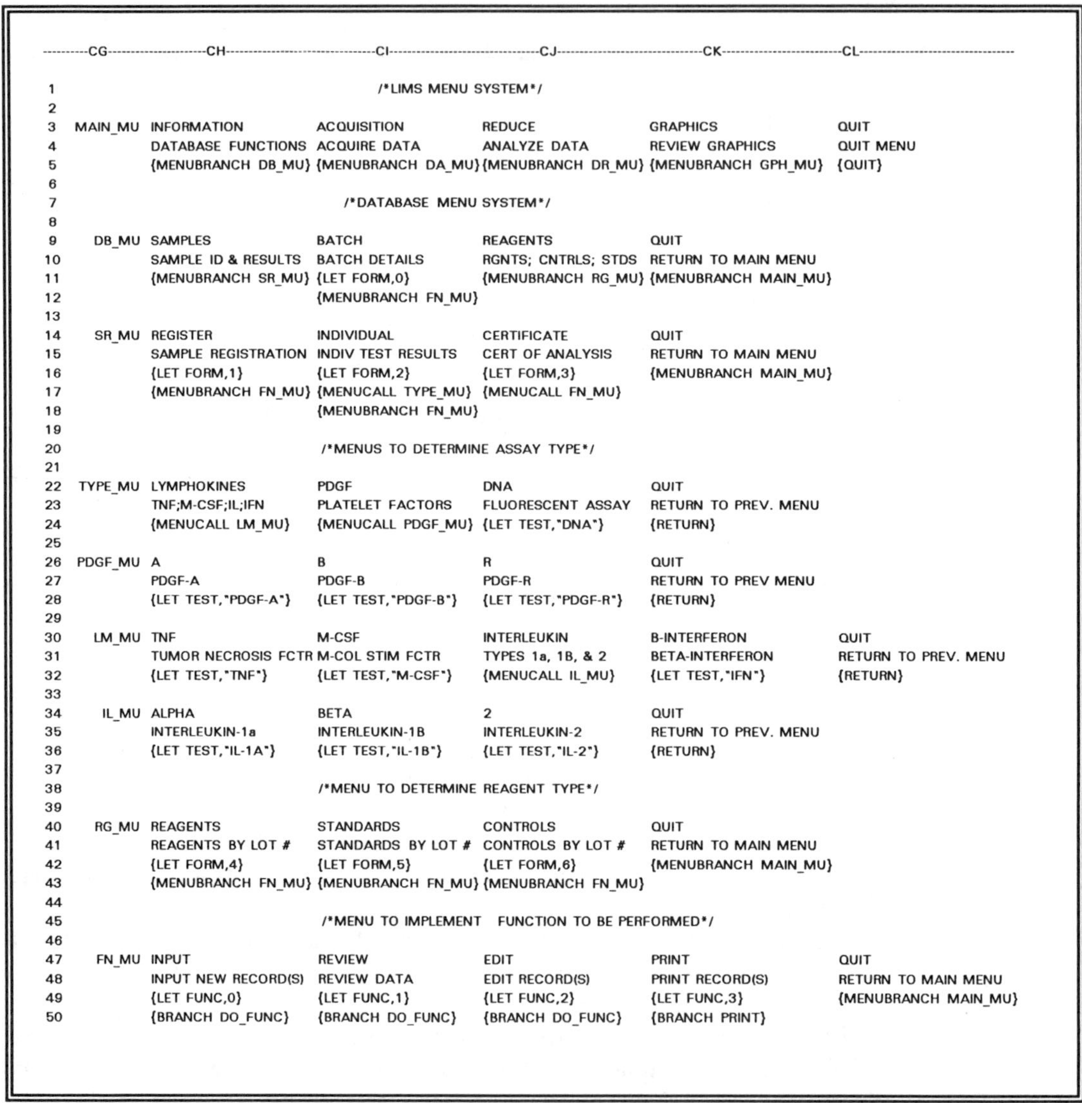

	CG	CH	CI	CJ	CK	CL
1			/*LIMS MENU SYSTEM*/			
2						
3	MAIN_MU	INFORMATION	ACQUISITION	REDUCE	GRAPHICS	QUIT
4		DATABASE FUNCTIONS	ACQUIRE DATA	ANALYZE DATA	REVIEW GRAPHICS	QUIT MENU
5		{MENUBRANCH DB_MU}	{MENUBRANCH DA_MU}	{MENUBRANCH DR_MU}	{MENUBRANCH GPH_MU}	{QUIT}
6						
7			/*DATABASE MENU SYSTEM*/			
8						
9	DB_MU	SAMPLES	BATCH	REAGENTS	QUIT	
10		SAMPLE ID & RESULTS	BATCH DETAILS	RGNTS; CNTRLS; STDS	RETURN TO MAIN MENU	
11		{MENUBRANCH SR_MU}	{LET FORM,0}	{MENUBRANCH RG_MU}	{MENUBRANCH MAIN_MU}	
12			{MENUBRANCH FN_MU}			
13						
14	SR_MU	REGISTER	INDIVIDUAL	CERTIFICATE	QUIT	
15		SAMPLE REGISTRATION	INDIV TEST RESULTS	CERT OF ANALYSIS	RETURN TO MAIN MENU	
16		{LET FORM,1}	{LET FORM,2}	{LET FORM,3}	{MENUBRANCH MAIN_MU}	
17		{MENUBRANCH FN_MU}	{MENUCALL TYPE_MU}	{MENUCALL FN_MU}		
18			{MENUBRANCH FN_MU}			
19						
20			/*MENUS TO DETERMINE ASSAY TYPE*/			
21						
22	TYPE_MU	LYMPHOKINES	PDGF	DNA	QUIT	
23		TNF;M-CSF;IL;IFN	PLATELET FACTORS	FLUORESCENT ASSAY	RETURN TO PREV. MENU	
24		{MENUCALL LM_MU}	{MENUCALL PDGF_MU}	{LET TEST,"DNA"}	{RETURN}	
25						
26	PDGF_MU	A	B	R	QUIT	
27		PDGF-A	PDGF-B	PDGF-R	RETURN TO PREV MENU	
28		{LET TEST,"PDGF-A"}	{LET TEST,"PDGF-B"}	{LET TEST,"PDGF-R"}	{RETURN}	
29						
30	LM_MU	TNF	M-CSF	INTERLEUKIN	B-INTERFERON	QUIT
31		TUMOR NECROSIS FCTR	M-COL STIM FCTR	TYPES 1a, 1B, & 2	BETA-INTERFERON	RETURN TO PREV. MENU
32		{LET TEST,"TNF"}	{LET TEST,"M-CSF"}	{MENUCALL IL_MU}	{LET TEST,"IFN"}	{RETURN}
33						
34	IL_MU	ALPHA	BETA	2	QUIT	
35		INTERLEUKIN-1a	INTERLEUKIN-1B	INTERLEUKIN-2	RETURN TO PREV. MENU	
36		{LET TEST,"IL-1A"}	{LET TEST,"IL-1B"}	{LET TEST,"IL-2"}	{RETURN}	
37						
38			/*MENU TO DETERMINE REAGENT TYPE*/			
39						
40	RG_MU	REAGENTS	STANDARDS	CONTROLS	QUIT	
41		REAGENTS BY LOT #	STANDARDS BY LOT #	CONTROLS BY LOT #	RETURN TO MAIN MENU	
42		{LET FORM,4}	{LET FORM,5}	{LET FORM,6}	{MENUBRANCH MAIN_MU}	
43		{MENUBRANCH FN_MU}	{MENUBRANCH FN_MU}	{MENUBRANCH FN_MU}		
44						
45			/*MENU TO IMPLEMENT FUNCTION TO BE PERFORMED*/			
46						
47	FN_MU	INPUT	REVIEW	EDIT	PRINT	QUIT
48		INPUT NEW RECORD(S)	REVIEW DATA	EDIT RECORD(S)	PRINT RECORD(S)	RETURN TO MAIN MENU
49		{LET FUNC,0}	{LET FUNC,1}	{LET FUNC,2}	{LET FUNC,3}	{MENUBRANCH MAIN_MU}
50		{BRANCH DO_FUNC}	{BRANCH DO_FUNC}	{BRANCH DO_FUNC}	{BRANCH PRINT}	

Figure 9-1

work. Along the way FORM and FUNC cells are updated with information pertaining to the template form to be used and type of function to be performed. At the bottom of the menu hierarchy in Figure 9-1, the program branches to programs called {DO_FUNC} and {PRINT}. These two programs are shown in Figure 9-2.

{DO_FUNC} serves as a distribution point that calls up the appropriate macro

```
--------HL-------HM-------HN-------HO-------HP----HQ-----HR-------HS-------HT---
1
2           /*PROGRAMS TO CALL FORM AND FUNCTION SPECIFIC MACROS*/
3
4           /*REVIEW AND EDIT DATA*/
5           /*NEEDS VALUES IN BOTH FORM AND FUNC CELLS*/
6  DO_FUNC  {IF FORM=0#AND#FUNC=1}{BRANCH IMP_BCH}  /*EXPORT BATCH*/
7           {LET SUB_PTR,@INDEX(FUNC_TAB,0,FORM)}
8           {DISPATCH SUB_PTR}/*BRANCH TO SUBROUTINE SPECIFIED IN SUB_PTR*/
9
10          /*PRINT THE CURRENT REPORT*/
11 PRINT    {LET PRN_SPEC,@INDEX(PRN_TAB,0,FORM)}
12          {WINDOWSOFF}{PANELOFF}   /*FREEZE THE DISPLAY*/
13          {IF FORM=0#AND#FUNC=0}{LET PRN_SPEC,"TEMPLATE"}
14          {IF RELEASE>4}{S}PSSR     /*SYMPHONY SPECIFIC PRINT COMMANDS*/
15          {IF RELEASE<4}/PPR        /*LOTUS 1-2-3 SPECIFIC PRINT COMMANDS*/
16 PRN_SPEC                           /*FORM TO PRINT*/
17          {IF RELEASE >4}~QAGPQ     /*SYMPONY SPECIFIC PRINT COMMANDS*/
18          {IF RELEASE <4}~AGPQ     /*LOTUS 1-2-3 SPECIFIC PRINT COMMANDS*/
19          {MENUBRANCH FN_MU}
20
21                    PRINT SPECIFICATION TABLE
22                            (PRN_TAB)
23                    ===========================
24                            BCH_FM          /*BATCH FORM*/
25                            SID_FM          /*SAMPLE ID FORM*/
26                            SRD_FM          /*SAMPLE RESULTS DETAIL FORM*/
27                            CERT_FM         /*CERTIFICATE OF ANALYSIS*/
28                            RG_FM           /*REAGENT FORM*/
29                            RG_FM           /*STANDARDS FORM*/
30                            RG_FM           /*CONTROLS FORM*/
31
32                    POINTER CELL SUBROUTINE TABLE
33                            (FUNC_TAB)
34                    =============================
35                            EXP_BCH
36                            DO_S_ID
37                            DO_SRD
38                            DO_CERT
39                            DO_RGT
40                            DO_STD
41                            DO_STD
42
43          /*MACRO TO BRING UP MAIN MENU*/
44        \M {MENUBRANCH MAIN_MU}
```

Figure 9-2

program for the form and function specified by the menu hierarchy. Different macro programs are required because {PUT} commands and @INDEX @functions need different range names, column offsets, and row offsets depending on the form they are interacting with. Beyond that requirement, the programs are very similar.

One method to accomplish the task of branching to the appropriate macro is to use a series of {IF} commands to test for each scenario and branch to the relevant macro to execute the task. The following is an example of this technique:

```
{IF FORM=0#AND#FUNC=1}{BRANCH IMP_BCH}
{IF FORM=0#AND#FUNC=0}{BRANCH EXP_BCH}
{IF FORM=1#AND#FUNC=1}{BRANCH DO_S_ID}
{IF FORM=2#AND#FUNC=1}{BRANCH DO_SRD}
{IF FORM=3#AND#FUNC=1}{BRANCH DO_CERT}
                   :
```

However, the method illustrated in Figure 9-2 is more straightforward, uses less memory, and is easier to expand if more forms are added to the system. The method is based on the {DISPATCH} command at the bottom of the macro. If you recall from Chapter 7, the {DISPATCH macro_spec} command causes a {BRANCH} to the macro that begins at the macro's name specified in the macro_spec cell and the new macro then begins executing.

In this instance, the macro_spec cell is called SUB_PTR. The {LET} command in cell HM7 places a macro name into SUB_PTR based on the form being used. The macro name is obtained from the Function Table (FUNC_TAB) in cells HO35..HO41 of Figure 9-2. The {LET} command uses an @INDEX to obtain the name from the table. The row offset in the @INDEX is the value in the FORM cell. Thus, the macro names in FUNC_TAB are arranged such that they correspond to the form being used. If FORM=3, then SUB_PTR would bear the string at row offset 3 of the table. That string is "DO_CERT". (Remember, the row offset in @INDEX @functions begins at zero, not one.)

The column offset in the @INDEX @function is set to zero because FUNC_TAB consists of a single column. Except for the {EXP_BCH} macro, all of the listed macros perform multiple purposes. They allow inputting, viewing, editing, and exporting of data. Because different forms are used for importing and exporting batch data, two different macro programs are needed. The two programs are {IMP_BCH} and {EXP_BCH}. The {IF} command at the beginning of {DO_FUNC} tests whether batch data are to be imported and branches to IMP_BCH when appropriate. If batch data are to be exported, the macro proceeds down to the {LET} command in cell HM7 and one of the macro names from the table are chosen.

If your system has several instances in which different macros are needed for different functions, you can expand the FUNC_TAB by adding more columns and placing the appropriate macro names into each column. You can then use an @INDEX @function of the following format to obtain the macro name for the SUB_PTR cell:

```
@INDEX(FUNC_TAB,FUNC,FORM)
```

The {PRINT} macro begins with a {LET} command that chooses a range name from the PRN_TAB (HO24..HO30) and places the range name of the form to be printed into the PRN_SPEC cell. The PRN_SPEC cell (HM16) is in the line of execution of the macro and supplies the name to the Lotus print settings.

Using the same logic as just described, a correction needs to be made when printing batch reports. (There are two batch report forms. See Figures 9-17 and 9-20.) The {IF} command in cell HM13 provides the amendment if necessary.

The pair of {IF} commands in cells HM14 and HM15 provide portability. If the Release number is greater than 4, Symphony is being used, and the [SERVICES] ***P***rint ***S***ettings ***S***ource ***R***ange command is issued. If the Release number is less than 4, Lotus 1-2-3 is being used, and the /***P***rint ***P***rinter ***R***ange command is issued.

The PRN_SPEC cell (HM16) is next encountered and the contents of the cell are automatically "typed" into the range settings. The tildes in cells HM17 and HM18 are equivalent to pressing [RETURN]. The final sequence is dependent on Lotus Release. For Symphony, the sequence is: ***Q***uit ***A***lign ***G***o ***P***age ***Q***uit. For Lotus 1-2-3, the sequence is: ***A***lign ***G***o ***P***age ***Q***uit.

{PRINT} concludes by branching back to the function menu. Without the {MENUBRANCH} command at the end of {PRINT}, the program would terminate and return to the spreadsheet.

The '\M macro at the bottom of Figure 9-2 is an important one to include in your programs. The '\M macro allows the user to start up the menu system by pressing [ALT]-M. Having a macro that will allow users to quit the menu system to explore data and later re-enter the menu system is user friendly. The macro is also essential during program development.

IMPORTING DATA FROM A DATABASE

Figure 9-3 illustrates two methods of importing data from a database and placing the data into a spreadsheet. The first method employs the {GET_FLD} subroutine described in Chapter 8 to import individual field data into the BUFFERA cell of the scratch-pad (Figure 9-21) and uses {PUT} commands to place the data into unordered cells of a template. The second method uses the {KEY_IMP} subroutine described in Chapter 8 to import field data into a rigidly-designed table. The first method is not as fast as the second, mainly because of the sluggish performance of the {PUT} commands. However, the first method provides more flexibility in data placement.

{DO_CERT} fills in the Certificate of Analysis form (Figure 9-4) with data from the SAMPLES.DBF and RESULTS.DBF databases. Actually, only the RESULTS.DBF database is accessed by the program. If you recall, RESULTS.DBF contains virtual fields that originate in the SAMPLES.DBF database and are a result of joining SAMPLES.DBF and RESULTS.DBF to form an extended relational database. The SAMPLES.DBF virtual fields are accessible through the RESULTS.DBF database.

The program begins with a {HOME} and {GOTO} set of commands. The {HOME} command positions the cell-pointer to cell A1. When the {GOTO} command re-positions the cell-pointer, the Certificate Of Analysis form will be centered. Without the {HOME} command, the form could take on different positions depending on whether the cell-pointer had been to the left or right of the CERT_FM range. The {HOME} command provides reliable centering of the form.

The {GETNUMBER} command pauses the macro and prompts the user for a

```
--------DA--------DB--------DC-------DD--------DE-------DF-------DG-------DH--
1                  /*FILL IN CERTIFICATE OF ANALYSIS FORM*/
2
3  DO_CERT  {HOME}{GOTO}CERT_FM~/*DISPLAY CERTIFICATION OF ANALYSIS FORM*/
4           {GETNUMBER "PLEASE ENTER SAMPLE'S BARCODE NUMBER: ",BCODE}
5           {PUT CERT_FM,3,5,BCODE} /*PUT BCODE INTO CERT OF ANALYSIS FORM*/
6           {PUT CERT_FM,1,1,@NOW}  /*PUT CURRENT DATE ON CERT OF ANALYSIS*/
7           {BLANK CERT_TAB}              /*BLANK RESULTS TABLE*/
8           {WINDOWSOFF}{PANELOFF}        /*FREEZE DISPLAY*/
9           {SEL_NDX "SAMPLES","SAMPL_BC"}/*SELECT JOIN PRIMARY INDEX*/
10          {SEL_NDX "RESULTS","RES_BC"}  /*SELECT RESULTS PRIMARY INDEX*/
11          {GET_REC "RESULTS",BCODE}     /*LOCATE RECORD*/
12          {PUT CERT_FM,6,1,RECRD}       /*PUT SAMPLE RECNO IN TEMPLATE*/
13          {FOR INC,0,5,1,PUT_SAMP}      /*PLACE RESULTS ON FORM*/
14          {IF @ISERR(RECRD)}{C_RNF}{BRANCH ABORT}/*PREVENT KEY_IMP ABORT*/
15          {KEY_IMP "RESULTS",@UPPER("CERT_HDR"),@STRING(BCODE,0)}
16          {BRANCH STOP}                 /*RETURN TO MENU SYSTEM*/
17
18          /*PUT DATA FROM A DATABASE FIELD INTO SPREADSHEET*/
19 PUT_SAMP {GET_FLD "RESULTS",@INDEX(C_TAB,0,INC),RECRD}/*RETURNS BUFFERA*/
20          {PUT CERT_FM,@INDEX(C_TAB,1,INC),@INDEX(C_TAB,2,INC),BUFFERA}
21
22          /*REPORT THAT RECORD IS NOT IN DATABASE*/
23 C_RNF    {PUT CERT_FM,3,6,"RECORD NOT FOUND"}
24
25                        CERT. OF ANALYSIS TABLE
26                                (C_TAB)
27                       ==============================
28                       REQ_NAME           2       8
29                       SAMP_NAME          3       6
30                       TYPE_REQ           3      12
31                       REQ_DT             2       9
32                       DEPT_NUM           6       8
33                       PROJ_CDE           6       9
34
```

Figure 9-3

barcode number. When [RETURN] is pressed, {GETNUMBER} places the barcode number into the BCODE cell. The {PUT} command in cell DB5 copies the barcode from BCODE to position (3,5) in the CERT_FM range. (The CERT_FM range is Y1..AF15.) Starting with Y1 as position (0,0), position (3,5) is cell AB6.

The {PUT} command in cell DB6 uses the @NOW @function to get current data and places the date into position (1,1) of the CERT_FM range.

Next, a {BLANK} command is used to erase any pre-existing data in the CERT_TAB range (Y20..AF30). When {KEY_IMP} is used, it places data into the CERT_TAB range. The next time {KEY_IMP} is used, it overwrites the data. However, if the second use of {KEY_IMP} retrieves fewer rows of data, some of the old data remain in the rows and appear as if they belonged to the current record being retrieved.

```
---------Y--------Z---------AA-------AB-------AC--------AD-------AE-----AF-
 1  CERT_FM 0,0
 2  DATE:        ________                        RECORD NUMBER:      _______
 3
 4                           CERTIFICATE OF ANALYSIS
 5
 6  SAMPLE BARCODE NUMBER:            _______
 7  SAMPLE NAME/DESCRIPTION:          _________________
 8
 9  REQUESTER'S NAME:         ________          DEPARTMENT NUMBER:  __
10  DATE OF REQUEST:          _______           PROJECT CODE:       _____
11
12
13  ASSAY TYPES REQUESTED:            ___________________________
14
15
16
17    ASSAY              DATE     BATCH    ASSAY     PLATE
18    TYPE     RESULTS PERFORMED NUMBER   DILUTION  NUMBER   STATUS  TECH
19  ASSAY_TYPE RESULTS  RUN_DT   BATCH  DIL_FCTR PLATE_NUMR_STATUS run_tech
20  _______  _______ _______ _______  _______ _______ _______ _______
21  _______  _______ _______ _______  _______ _______ _______ _______
22  _______  _______ _______ _______  _______ _______ _______ _______
23  _______  _______ _______ _______  _______ _______ _______ _______
```

Figure 9-4

The {BLANK} command erases all old data and ensures that only current data are in the CERT_TAB range.

Next, {SEL_NDX} is called to select the primary indexes for the operation. As explained in Chapter 8, if there are several open index files for a database file, you must specify which index file is primary. The primary index is the index that contains the data and index key expression that are used to locate records in the database. If you use the wrong index file as the primary, searches will be ineffectual. Therefore, you should use {SEL_NDX} to set the primary indexes in all programs that you develop, whether you think the action is needed or not.

In this example, SAMPL_BC and RES_BC are selected as primary index files for the SAMPLES.DBF and RESULTS.DBF databases, respectively. Therefore, searches by BARCODE are being targeted.

After selecting the primary indexes, the {GET_REC} subroutine from Chapter 8 is called to determine the number of the first record matching the value in the BCODE cell. Finding a record number is required before any data can be retrieved from a database. The {FOR} command in cell DB13 repeatedly calls {PUT_SAMP}. {PUT_SAMP} uses the record number and an @INDEX @function to retrieve field data from the RESULTS.DBF database. {PUT_SAMP} is executed 6 times (0 through 5). Each time {PUT_SAMP} is called, the @INDEX retrieves a field name from the C_TAB range (DC28..DE33). For example, the first time {PUT_SAMP} is executed, INC=0 and the field name is REQ_NAME. The second time {PUT_SAMP} is executed, INC=1, the field name is SAMP_NAME, and so on.

{GET_FLD} returns field information for the record number. The {PUT} command transmits the information to CERT_FM. The two @INDEX @functions in the {PUT} command direct the coordinates to be used. These @INDEX @functions obtain the column and row coordinates from columns 1 and 2 of the C_TAB table. For example, the first time {PUT_SAMP} is executed, INC=0 and the row and column coordinates are 2 and 8, respectively.

After the field information has been placed into the template a test is made to ensure that a record exists for BCODE. The {GET_FLD} subroutine will return ERRs to all cells if a record does not exist and the program will continue to execute. However, {KEY_IMP} is based on @BASE menu commands. If a record does not exist and you try to use {KEY_IMP}, the program would terminate and an error message would be displayed. To prevent this action, the {IF} command in cell DB14 tests whether {GET_REC} placed an ERR in the RECRD cell. Placing the {IF} test after the {FOR} loop allows the {FOR} loop to fill the template with ERRs so that the user notices that the record does not exist.

If the RECRD cell has an ERR in it, the {C_RNF} subroutine is called (Certificate-Record Not Found). {C_RNF} places the "Record Not Found" string into the SAMPLE NAME/DESCRIPTION cell to further indicate the absence of the sample. The macro then branches to {ABORT} to return to FN_MU. This {BRANCH} eliminates the macro termination problem associated with {KEY_IMP}.

If a record exists, {KEY_IMP} is called to import tabular data and append the data to the bottom of the CERT_FM template. {KEY_IMP} uses labels in the CERT_HDR cells (Y19..AF19) as specifications for the fields to import. All records in the RESULTS.DBF database matching the BCODE key value are imported.

The @UPPER @function in the argument list of {KEY_IMP} deserves an explanation. If you recall from Chapter 8, it is crucially important to observe argument type when passing arguments to subroutines. The IMP_LBLS argument of {KEY_IMP} was defined with a ":VALUE" suffix. Therefore, the argument passed for IMP_LBLS must be a cell address or formula. The @UPPER makes no substantive change on the string, but fulfills the requirement because @INDEX is considered to be a formula.

A word of warning needs to be reiterated on the use of {KEY_IMP}. {KEY_IMP} overwrites everything in cells that fall within the output range. Be sure to leave enough space below the row of import labels to account for the maximum number of records expected to be imported.

EXPORTING DATA TO A DATABASE

The {DO_STD} program (Figures 9-5 and 9-6) is similar to {DO_CERT}, but adds exporting features. {DO_STD} supports the standards and controls database functions of the LIM system. The form supported by {DO_CERT} is a named range called STD_FM and is illustrated in Figure 9-7.

```
--------FR-------FS-------FT-------FU--------FV--------FW-------FX------FY---
1
2          /*PROGRAM TO TRANSFER STANDARD OR CONTROL DATA*/
3          /*BETWEEN STANDARD.DBF DATABASE AND STD_FM TEMPLATE*/
4          /*(STD_FM AND STANDARD.DBF SERVE DUAL PURPOSES)*/
5
6  DO_STD  {IF FORM=5}{PUT STD_FM,3,3,"   STANDARDS INFORMATION"}
7          {IF FORM=6}{PUT STD_FM,3,3,"    CONTROLS INFORMATION"}
8          {HOME}{GOTO}STD_FM~          /*DISPLAY STANDARDS FORM*/
9          {SEL_NDX "STANDARD","STDS"} /*SPECIFY INDEX FILE TO USE*/
10         {GETLABEL "PLEASE ENTER LOT NUMBER: ",BUFFERA}
11         {PAD "STANDARD","LOT_STD",BUFFERA}    /*CNVERT TO IKE SPEC*/
12         {PUT STD_FM,3,5,BUFFERA}     /*PUT LOT NUMBER INTO TEMPLATE*/
13         {PUT STD_FM,1,1,@NOW}        /*PUT CURRENT DATE INTO FORM*/
14         {GET_REC "STANDARD",BUFFERA}/*SEE IF RECORD EXISTS; GET RECNO*/
15         {IF RECRD<>0}{FOR INC,0,21,1,PUT_STD} /*ADD PRIOR DATA*/
16         {IF @ISERR(RECRD)}{PUT STD_FM,3,7,"RECORD NOT FOUND"}
17         {CALC}              /*FORCE UPDATE OF DISPLAY*/
18         {IF FUNC=1}{BRANCH STOP}     /*REVIEW ONLY*/
19         {IF FUNC=3}{BRANCH PRINT}    /*PRINT THE REAGENT FORM*/
20         {LET SUB_PTR,"STD_EXP"} /*NEEDED FOR {DISPATCH} IN {USR_INP}*/
21         {BRANCH USR_INP}             /*ALLOW USER TO FILL/EDIT FORM*/
22
23         /*EXPORT STANDARD OR CONTROL DATA TO REAGENTS.DBF*/
24 STD_EXP {LET BUFFERA,@INDEX(STD_FM,3,5)}       |*RECHECK TO SEE IF    *|
25         {MK_STRNG BUFFERA}                     |*USER CHANGED LOT NUMB*|
26         {PAD "STANDARD","LOT_STD",BUFFERA}    /*CNVERT TO IKE SPEC*/
27         {CHK_CRT "STANDARD","LOT_STD",BUFFERA}/*LOOK FOR RCRD*/
28         {PUT STD_FM,9,1,RECRD}       /*PUT RECNO INTO TEMPLATE*/
29         {MK_DT @INDEX(STD_FM,3,11)} /*CONVERT PREPARATION DATE*/
30         {PUT STD_FM,3,11,BUFFERA}    /*PLACE DATE INTO TEMPLATE*/
31         {FOR INC,0,13,1,SND_STDA}    /*UPDATE DATABASE*/
32         {FOR INC,14,21,1,SND_STDB}   /*CONVERT LOT NUMS AND TRANSFER*/
33         {BRANCH STOP}      /*UN-FREEZE DISPLAY, RETURN TO FN_MU*/
34
```

Figure 9-5

{DO_CERT} begins by determining the value in the FORM cell. The value of the FORM cell is set by a Standards or Controls choice in the RG_MU menu of Figure 9-1. By determining the value of FORM, the template is modified by placing FORM-specific titles into the template. This modification allows the template to be used for both functions, thereby conserving memory and programming time.

After centering the display on STD_FM and selecting the STDS.NDX index file as the primary index, the user is prompted for the lot number. A {GETLABEL} command is used for this task. If you recall from the REAGENTS.DBF database definition (Figure 2-7), LOT_STD was defined as a character string. {GETLABEL} converts any input to character strings, eliminating the need for users to be concerned with numbers vs. character strings. The string is placed into BUFFERA and padded to conform to the index key expression used in STDS.NDX.

```
--------FR-------FS-------FT-------FU--------FV--------FW-------FX------FY---
35          /*UTILITIES FOR IMPORTING AND EXPORTING STANDARDS DATA*/
36
37          /*ADD EACH FIELD OF DATA TO TEMPLATE*/
38 PUT_STD  {GET_FLD "STANDARD",@INDEX(STD_TAB,0,INC),RECRD}
39          {LET BUFFERB,@INDEX(STD_TAB,1,INC)}   /*MAKE COLUMN COORD*/
40          {LET BUFFERC,@INDEX(STD_TAB,2,INC)}   /*MAKE ROW COORD*/
41          {PUT STD_FM,BUFFERB,BUFFERC,BUFFERA}/*PUT DATA INTO TEMPLATE*/
42
43          /*EXPORT STANDARDS DATA TO DATABASE*/
44 SND_STDA {ASMBL_STD}        /*ASSEMBLE VALUE FROM TEMPLATE*/
45          {SND_STDC}         /*TRANSFER VALUE TO DATABASE*/
46
47          /*ENSURE LOT NUMBERS ARE STRINGS AND EXPORT TO DBASE*/
48 SND_STDB {ASMBL_STD}        /*ASSEMBLE VALUE FROM TEMPLATE*/
49          {MK_STRNG BUFFERA}/*CONVERT LOT NUMBERS TO STRINGS*/
50          {SND_STDC}         /*TRANSFER VALUE TO DATABASE*/
51
52          /*ASSEMBLE A VALUE FROM STD_FM*/
53 ASMBL_STD{LET BUFFERB,@INDEX(STD_TAB,1,INC)}   /*COLUMN OFFSET*/
54          {LET BUFFERC,@INDEX(STD_TAB,2,INC)}   /*ROW OFFSET*/
55          {LET BUFFERA,@INDEX(STD_FM,BUFFERB,BUFFERC)}/*VALUE FROM FORM*/
56
57          /*TRANSFER A VALUE FROM TEMPLATE TO DATABASE RECORD*/
58 SND_STDC {UPDATE "STANDARD",@INDEX(STD_TAB,0,INC),BUFFERA}
59
60                    STANDARD TABLE
61                      (STD_TAB)
62          ============================
63          LOT_STD          3         5
64          STD_NAME         3         7
65          STD_DESC         3         8
66          S_PREP_DT        3        10
67          S_SHELF_L        3        11
68          STD_TECH         3        12
69          CONC_S0          4        15
70          CONC_S1          4        16
71          CONC_S2          4        17
72          CONC_S3          4        18
73          CONC_S4         10        15
74          CONC_S5         10        16
75          CONC_S6         10        17
76          CONC_S7         10        18
77          LOT_S0           1        15
78          LOT_S1           1        16
79          LOT_S2           1        17
80          LOT_S3           1        18
81          LOT_S4           7        15
82          LOT_S5           7        16
83          LOT_S6           7        17
84          LOT_S7           7        18
```

Figure 9-6

```
-------BF------BG----BH----BI-----BJ-BK--BL-------BM----BN----BO-------BP---BQ
  1  0,0 STD_FM
  2  DATE:________                          RECORD NUMBER:      ________
  3
  4                                STANDARDS INFORMATION
  5
  6       LOT NUMBER:   ________
  7
  8      NAME OF SET:   ________
  9  SET DESCRIPTION:   ________
 10
 11  PREPARATION DATE:  ________              EXPIRATION DATE:________
 12  SHELF LIFE:        ________              EXPIRATION FLAG:________
 13  TECHNICIAN:        ________
 14
 15             LOT NUMBER         [CONC]            LOT NUMBER          [CONC]
 16      0 ____________________   _______    4 ____________________   ________
 17      1 ____________________   _______    5 ____________________   ________
 18      2 ____________________   _______    6 ____________________   ________
 19      3 ____________________   _______    7 ____________________   ________
 20           (PRESS [CONTROL][BREAK] WHEN DONE ENTERING DATA)
```

Figure 9-7

{GET_REC} is then called to determine whether a record exists for the lot number. If a record exists, existing field data are placed into the template. This action provides some useful features:

- If data exist, a user does not have to re-enter the data from the start.
- The form can be used for multiple purposes: entering, editing, and printing data.
- The user immediately knows that an existing record exists so that pre-existing data are not inadvertently overwritten.

The {IF} command in cell FS15 tests whether the record exists. If the record exists, field information is placed into the template using a {FOR} loop set to repeatedly call {PUT_STD}. {PUT_STD} would be similar in form and function to {PUT_SAMP} in Figure 9-3. An alternative strategy is to remove the {IF} test and allow {PUT_SAMP} to place ERRs into the cells if a record did not exist. These ERRs would alert a user that a record is absent. The ERRs would also mark positions where data needed to be entered.

The {IF} command in cell FS18 determines whether the FUNC cell is one. If you recall from the FN_MU in Figure 9-1, a one is placed into the FUNC cell if the form is to be used for review only.

Next "STD_EXP" is placed into the SUB_PTR cell. The information in the SUB_PTR cell will be used later by a {DISPATCH} command to specify a {BRANCH} to {STD_EXP}.

A {BRANCH} is next made to {USR_INP} to allow the user to fill in the STD_FM

```
--------GO-------GP-------GQ--------GR--------GS-------GT-------GU-------GV---
 1
 2          /*PROGRAM TO TRANSFER SAMPLE IDENTIFICATION DATA*/
 3          /*BETWEEN SAMPLES.DBF DATABASE AND SID_FM TEMPLATE*/
 4
 5  DO_S_ID {HOME}{GOTO}SID_FM~       /*GO TO SAMPLE ID FORM*/
 6          {SEL_NDX "SAMPLES","SAMPL_BC"}/*SPECIFY INDEX FILE TO USE*/
 7          {GETNUMBER "PLEASE ENTER SAMPLE BARCODE NUMBER: ",BUFFERA}
 8          {PUT SID_FM,3,5,BUFFERA}  /*PUT BARCODE NUMBER INTO FORM*/
 9          {PUT SID_FM,1,1,@NOW}     /*PLACE TODAY'S DATE INTO FORM*/
10          {GET_REC "SAMPLES",BUFFERA}/*SEE IF RECORD EXISTS*/
11          {IF RECRD<>0}{FOR INC,0,11,1,GET_SID} /*GET EXISTING DATA*/
12          {CALC}                    /*FORCE UPDATE OF DISPLAY*/
13          {IF FUNC=1}{BRANCH STOP}  /*DATA IS FOR DISPLAY ONLY*/
14          {IF FUNC=3}{BRANCH PRN_IT}/*PRINT THE SAMPLE ID FORM*/
15          {LET SUB_PTR,"EXP_SID"}   /*NEEDED TO RETURN FROM {USR_INP}*/
16          {BRANCH USR_INP}          /*ALLOW USER INPUT FOR ENTRY FORM*/
17
18          /*EXPORT SAMPLE IDENTIFICATION DATA TO SAMPLES.DBF*/
19  EXP_SID {CHK_CRT "SAMPLES","BARCODE",@INDEX(SID_FM,3,5)}
20          {PUT SID_FM,6,1,RECRD}    /*PUT RECORD NUMBER INTO SID_FM*/
21          {MK_DT @INDEX(SID_FM,2,8)}/*CONVERT DATE OF REQUEST*/
22          {PUT SID_FM,2,8,BUFFERA}  /*PUT DATE INTO FORM*/
23          {MK_DT @INDEX(SID_FM,6,8)}/*CONVERT SAMPLE DELIVERY DATE*/
24          {PUT SID_FM,6,8,BUFFERA}  /*PUT DATE INTO FORM*/
25          {FOR INC,0,10,1,PUT_SID}  /*TRANSFER DATA TO DATABASE*/
26          {BRANCH STOP}    /*UN-FREEZE DISPLAY, RETURN TO FN_MU*/
27
28          /*PUT SAMPLE ID DATA INTO SAMPLES.DBF*/
29  PUT_SID {LET BUFFERB,@INDEX(SI_TAB,1,INC)}   /*COLUMN COORD*/
30          {LET BUFFERC,@INDEX(SI_TAB,2,INC)}   /*ROW COORD*/
31          {LET BUFFERD,@INDEX(SI_TAB,0,INC)}   /*FIELD NAME TO UPDATE*/
32          {UPDATE "SAMPLES",BUFFERD,@INDEX(SID_FM,BUFFERB,BUFFERC)}
33
34          /*GET SAMPLE ID DATA FROM SAMPLES.DBF*/
35  GET_SID {GET_FLD "SAMPLES",@INDEX(SI_TAB,0,INC),RECRD}
36          {PUT SID_FM,@INDEX(SI_TAB,1,INC),@INDEX(SI_TAB,2,INC),BUFFERA}
37
38                   SAMPLE IDENTIFICATION TABLE
39                            (SI_TAB)
40                   ==============================
41                   BARCODE            3       5
42                   SAMP_NAME          3       6
43                   REQ_DT             2       8
44                   REQ_NAME           2       9
45                   DEPT_NUM           2      10
46                   PROJ_CDE           2      11
47                   SMP_DEL_DT         6       8
48                   RACK_BCD           6       9
49                   STOR_LOC           6      10
50                   TYPE_REQ           2      13
51                   REQ_NTS            2      14
52                   S_STATUS           4      16
```

Figure 9-8

```
--------O---------P---------Q---------R--------S--------T--------U----
 1  SID_FM 0,0
 2  DATE:          ________                      RECORD NUMBER:     _______
 3
 4                          SAMPLE IDENTIFICATION
 5
 6  SAMPLE BARCODE NUMBER:             _______
 7  SAMPLE NAME/DESCRIPTION:           __________________
 8
 9  DATE OF REQUEST:       _______      SAMPLE DELIVERY DATE:      ________
10  REQUESTER'S NAME:      _________    SAMPLE RACK BARCODE:       _______
11  DEPARTMENT NUMBER:     ___          STORAGE LOCATION:          ________
12  PROJECT CODE:          _____
13
14  ASSAYS REQUESTED:      _____________________________
15  COMMENTS:              ___________________________
16
17                         SAMPLE STATUS:        ___
18
19      (PRESS [CONTROL][BREAK] WHEN DONE ENTERING DATA)
20
```

Figure 9-9

template. After [CONTROL] and [BREAK] are pressed, a {DISPATCH} command in {USR_INP} makes a branch to {STD_EXP} to export the reagent data. {STD_EXP} first checks the lot number to ensure that a number was not placed into cell BI6. A number cannot be used when searching the index file. LOT_STD must be a string. If the lot number is a number, {MK_STRNG} converts it to a string and then {PAD} pads the lot number for the appropriate number of spaces in the index key expression. {CHK_CRT} is then called to see if the record exists. Calling {CHK_CRT} fulfills two goals:

- If the record exists, {CHK_CRT} will return its record number to BUFFERA.
- If the record does not exist, {CHK_CRT} will create a record, place the lot number in cell BI6 into the LOT_STD field, and return the record number.

Next, {MK_DT} is called to convert the preparation date to a Lotus serial date number and the numeric value of cell BI11 of Figure 9-7 is returned to the form. Then, {SND_STDA} and {SND_STDB} (Figure 9-5) are called repeatedly to export the data to the database fields. The difference in these two subroutines is the presence of a call to {MK_STRNG} in {SND_STDB} to convert lot numbers to strings.

The {ASMBL_STD} subroutine assembles column and row offset data from STD_TAB (FS63..FU84) into BUFFERB and BUFFERC and uses the offset data to retrieve data from the STD_FM template. The @INDEX in cell FS55 uses these offsets to pinpoint the data to be retrieved from the template. The {SND_STDC} updates the database using the {UPDATE} subroutine described in Chapter 8. The field name used each time {UPDATE} is called is selected from the STD_TAB using an @INDEX and

```
--------GD------GE-------GF--------GG---------GH-------GI-------GJ-------GK-------
 1
 2           /*PROGRAM TO GET INDIVIDUAL ASSAY DATA ON A SAMPLE*/
 3
 4  DO_SRD   {HOME}{GOTO}SRD_FM~ /*DISPLAY CERTIFICATE OF ANALYSIS FORM*/
 5           {GETNUMBER "PLEASE ENTER SAMPLE'S BARCODE NUMBER: ",BCODE}
 6           {PUT SRD_FM,3,5,BCODE}  /*PUT BARCODE INTO SAMPLE RESULTS DETAIL FORM*/
 7           {PUT SRD_FM,1,1,@NOW}   /*PUT CURRENT DATE ON CERT OF ANALYSIS FORM*/
 8           {SEL_NDX "SAMPLES","SAMPL_BC"}  /*SELECT JOIN PRIMARY INDEX*/
 9           {SEL_NDX "RESULTS","RES_BC"}    /*SELECT RESULTS PRIMARY INDEX*/
10           {GET_REC "RESULTS",BCODE}       /*LOCATE RECORD*/
11           {WINDOWSON}{PANELON}            /*UN-FREEZE DISPLAY*/
12           {PUT_SRD}                       /*PUT RESULTS INTO TEMPLATE*/
13           {LET SUB_PTR,"DO_SRD"}          /*NEEDED FOR {DISPATCH} IN NPD_MU*/
14           {MENUBRANCH NPD_MU}     /*SEE IF NEXT OR PREVIOUS RECORD WANTED*/
15
16           /*RECORD NOT FOUND*/
17  SRD_NF   {PUT SRD_FM,3,6,"RECORD NOT FOUND"}
18
19           /*GET NEXT SAMPLE RESULTS RECORD*/
20  NEXT_SRD {GET_NEXT "RESULTS",BCODE}
21           {PUT_SRD}             /*PUT RESULTS INTO TEMPLATE*/
22           {MENUBRANCH NPD_MU} /*BRANCH BACK TO NEXT/PREVIOUS MENU*/
23
24           /*GET PREVIOUS SAMPLE RESULTS RECORD*/
25  PREV_SRD {GET_PREV "RESULTS",BCODE}
26           {PUT_SRD}             /*PUT RESULTS INTO TEMPLATE*/
27           {MENUBRANCH NPD_MU} /*BRANCH BACK TO NEXT/PREVIOUS MENU*/
28
29           /*PUT RESULTS INTO TEMPLATE*/
30  PUT_SRD  {PUT SRD_FM,6,1,RECRD}          /*PUT RESULTS RECNO IN TEMPLATE*/
31           {FOR INC,0,21,1,IMP_SRD}        /*PLACE RESULTS INTO FORM*/
32           {IF @ISERR(RECRD)}{SRD_NF}      /*SIGNAL RECORD NOT FOUND*/
33           {CALC}                          /*FORCE UPDATE OF SPREADSHEET*/
34
35           /*IMPORT DATA FROM RESULTS.DBF AND PLACE INTO TEMPLATE*/
36  IMP_SRD  {GET_FLD "RESULTS",@INDEX(SRD_TAB,0,INC),RECRD}
37           {PUT SRD_FM,@INDEX(SRD_TAB,1,INC),@INDEX(SRD_TAB,2,INC),BUFFERA}
38
```

Figure 9-10

the current value of INC.

The {DO_S_ID} program in Figure 9-8 is similar to {DO_STD}. The {DO_S_ID} program transfers data between the SAMPLES.DBF database file and the SID_FM template (Figure 9-9). It would be good practice to review each section of {DO_S_ID} to compare the program to {DO_STD}. Note especially, the absence of a call to the {PAD} subroutine. If you recall, BARCODE is a number. Using {PAD} is neither warranted, nor wanted, when using numbers with indexes that have index key expressions

```
--------GD-------GE--------GF---------GG---------GH-------GI-------GJ------
38
39                      SAMPLE RESULTS DETAIL TABLE
40                                 (SRD_TAB)
41                      ================================
42                      SAMP_NAME              3       6
43                      REQ_NAME               2       7
44                      REQ_DT                 2       8
45                      DEPT_NUM               6       7
46                      PROJ_CDE               6       8
47                      ASSAY_TYPE             2      10
48                      RESULTS                2      11
49                      DIL_FCTR               6      10
50                      RUN_DT                 2      14
51                      BATCH                  2      15
52                      RAW_DATA1              2      16
53                      RAW_DATA2              2      17
54                      PLATE_NUM              6      14
55                      BCH_TECH               6      15
56                      LOT_STD                6      16
57                      LOT_RGNT               6      17
58                      CURV_A                 2      20
59                      STD_Y                  2      21
60                      STD_C                  2      22
61                      CURV_B                 5      20
62                      R_SQR                  5      21
63                      TECH_COMS              2      24
```

Figure 9-11

that are based on single numerical fields.

GETTING THE NEXT OR PREVIOUS RECORD

Figures 9-10 through 9-12 illustrate how to design a program and template that provide users with the capability of stepping forward and backward through a series of records having the same key value. An example situation would be one in which samples had several different assay types performed on them and the index key expression was based on barcodes. The RESULTS.DBF database file may then have several records for each barcode. If the {GET_REC} subroutine was called to get the record number, only the first occurrence of the barcode would be retrieved. To access the other records, you would need to use the {GET_NEXT} and {GET_PREV} subroutines described in Chapter 8.

The {DO_SRD} program illustrates how to use these subroutines. {DO_SRD} is similar in form and function to previous programs in this chapter. The display is aligned, user input obtained, primary index selected, etc. The process of obtaining sequential records begins with a call to the {GET_REC} subroutine in cell GE10. Before

```
--------AK-------AL-------AM-------AN-------AO-------AP-------AQ---
 1  SRD_FM 0,0
 2  DATE:        _________                    RECORD NUMBER:     _________
 3
 4                          SAMPLE RESULTS DETAIL
 5
 6  SAMPLE BARCODE NUMBER:           _________
 7  SAMPLE NAME/DESCRIPTION:         __________________
 8  REQUESTER'S NAME:  _________              DEPARTMENT NUMBER:_________
 9  DATE OF REQUEST:   _________              PROJECT CODE:      _________
10
11  ASSAY TYPE:        _________              DILUTION FACTOR:   _________
12  RESULTS:           _________
13
14                    *****BATCH DETAILS*****
15  RUN DATE:          _________              PLATE NUMBER:      _________
16  BATCH:             _________              TECHNICIAN:        _________
17  DUPLICATE #1:      _________              STD LOT NUMBER:    _________
18  DUPLICATE #2:      _________              RGT LOT NUMBER:    _________
19
20                  ****STANDARD CURVE DETAILS****
21           SLOPE:    ________ INTERCEPT:          ________
22   STD ERR Y EST:    ________ R SQUARED:          ________
23    STD ERR COEF:    ________
24
25       COMMENTS:______________________________
26
```

Figure 9-12

{GET_NEXT} and {GET_PREV} can be used to move back and forth through records, you must first use the {GET_REC} subroutine to locate the first occurrence of the record matching the search key.

The {LET} command in cell GE13 places "DO_SRD" into the SUB_PTR cell. This action is required because the Different menu choice on the NPD_MU (Figure 8-2) allows a user to review records for different BARCODE numbers. Because NPD_MU uses a {DISPATCH} command at the bottom of the Different menu choice to {BRANCH} to the next program, a {LET} command must specify "DO_SRD". This method allows the NPD_MU menu to be generic.

The {MENUBRANCH NPD_MU} command throughout the program brings up the NPD_MU menu after the SRD_FM template is filled with data. This command and the resulting menu allow users to select the next record to be displayed.

It would be good practice to add the capability to find the first and last occurrences of records into the RESULTS.DBF database. The following is a list of tasks that you will need to complete to add the capability:

- Create a generic subroutine utility named {GET_LAST}. {GET_LAST} should

```
--------GZ-------HA-------HB---------HC---------HD-------HE-------HF-------
 1          /*PROGRAM TO IMPORT BATCH DATA FROM DATABASE*/
 2                  /*INTO BATCH DETAIL TEMPLATE*/
 3
 4  IMP_BCH  {HOME}{GOTO}BCH_FM~              /*DISPLAY BATCH DETAIL FORM*/
 5           {LET FORM,0}{LET FUNC,1}         /*SET FORM AND FUNCTION*/
 6           {WINDOWSOFF}{PANELOFF}           /*FREEZE THE DISPLAY*/
 7           {MENUCALL TRIO_MU} /*GET ASSAY TYPE, RUN DATE, & BATCH NUMBER*/
 8           {PUT BCH_FM,1,1,@NOW}        /*PUT CURRENT DATE INTO BATCH FORM*/
 9           {BLANK BCH_TAB}                  /*BLANK SAMPLE RESULTS TABLE*/
10           {WINDOWSOFF}{PANELOFF}           /*FREEZE DISPLAY*/
11           {SEL_NDX "BATCH","BCH_TRIO"}     /*SELECT PRIMARY INDEX*/
12           {SEL_NDX "RESULTS","RES_TRIO"}   /*SELECT PRIMARY INDEX*/
13           {GET_REC "BATCH",KEY}            /*LOCATE RECORD*/
14           {PUT BCH_FM,6,1,RECRD}           /*PUT BATCH RECNO IN TEMPLATE*/
15           {FOR INC,0,23,1,PUT_BCH}         /*PLACE RESULTS ON FORM*/
16           {GET_REC "RESULTS",KEY}          /*SEE IF ANY SAMPLES PRESENT*/
17           {IF @ISERR(RECRD)}{B_RNF}{BRANCH ABORT}
18           {KEY_IMP "RESULTS",@UPPER("BCH_HDR"),BUFFERD}
19           {BRANCH STOP}                    /*RETURN TO MENU SYSTEM*/
20
21           /*PUT DATA FROM A DATABASE FIELD INTO SPREADSHEET*/
22  PUT_BCH  {GET_FLD "BATCH",@INDEX(BD_TAB,0,INC),RECRD} /*RETURNS BUFFERA*/
23           {PUT BCH_FM,@INDEX(BD_TAB,1,INC),@INDEX(BD_TAB,2,INC),BUFFERA}
24
25           /*REPORT THAT RECORD IS NOT IN DATABASE*/
26  B_RNF    {PUT BCH_FM,2,8,"RECORD NOT FOUND"}
27
```

Figure 9-13

be similar in format to {GET_REC}, but based on the @BASE @NDXLAST @function. @NDXLAST returns the record number of the last occurrence of the key value in an index file. If a record cannot be found, a zero is returned. The syntax of @NDXLAST is:

```
@NDXLAST(DATBASE ALIAS,SEARCH VALUE)
```

@NDXLAST is similar to using @DBLAST, but is much faster because it uses the primary index to perform the lookup.

- Create subroutines called {FIRST_SRD} and {LAST_SRD}. These subroutines should be similar in form and function to {NEXT_SRD} and {PREV_SRD}. However, they should call {GET_REC} and {GET_LAST} instead of {GET_NEXT} and {GET_PREV}.
- Move the Different and Quit menu choices in the NPD_MU of Figure 8-2 from cells IB9..IC11 to cells IE9..IF11.
- Create menu choices for First and Last in columns IC and ID, respectively. At the bottom of these menu choices, use {BRANCH} commands to branch to

```
--------GZ-------HA-------HB---------HC---------HD-------HE-------HF-------
28
29                     BATCH DETAIL OFFSET TABLE
30                              (BD_TAB)
31                    ================================
32                    ASSAY_TYPE          2       5
33                    RUN_DT              2       6
34                    LOT_RGNT            2       8
35                    LOT_STD             2       9
36                    BATCH               6       5
37                    BCH_TECH            6       6
38                    CONC_S0             0      12
39                    CONC_S1             0      13
40                    CONC_S2             0      14
41                    CONC_S3             3      12
42                    CONC_S4             3      13
43                    CONC_S5             3      14
44                    AVG_S0              1      12
45                    AVG_S1              1      13
46                    AVG_S2              1      14
47                    AVG_S3              4      12
48                    AVG_S4              4      13
49                    AVG_S5              4      14
50                    CURV_A              1      18
51                    CURV_B              1      17
52                    TECH_COMS           1      20
53                    STD_Y               6      17
54                    R_SQR               6      18
55                    FREV_DTE            6      20
56
```

Figure 9-14

{FIRST_SRD} and {LAST_SRD}, respectively.

USING INDEX KEY EXPRESSIONS WITH MULTIPLE FIELD NAMES

All of the previously documented programs dealt with index key expressions based on single field names. Different techniques are needed for searching index files based on two or more fields. With multiple keys, a key value must be prepared by concatenating strings containing the specified fields. Before strings are concatenated, they must be padded using the {PAD} subroutine. One of the most effective methods for obtaining multiple key values from a user is to use a menu system. Figures 9-13 through 9-16 illustrate a program and menu system to import batch data from the BATCH.DBF database into the BCH_FM template shown in Figure 9-17. Figures 9-18 and 9-19 show a program for exporting batch data from the template shown in Figure 9-20 to the BATCH.DBF database file.

The {IMP_BCH} program in Figure 9-13 is similar to the previous programs in this chapter. The only substantive difference is in the way that the key value is handled for

```
--------EV---------EW---------------EX-----------------EY-----------------EZ--------
1
2            /*UTILITY PROGRAMS FOR INDEXES CONTAINING THREE KEYS*/
3
4            /*MENU TO GET ASSAY_TYPE/RUN_DT/BATCH TRIO*/
5 TRIO_MU   CURRENT      TYPE                DATE                BATCH
6           USE TEMPLATE CHANGE ASSAY TYPE   CHANGE ASSAY DATE   CHANGE BATCH #
7           {MK_TRIO}    {GET_TYPE}          {GET_DT}            {GET_BNUM}
8                        {MENUBRANCH TRIO_MU}{MENUBRANCH TRIO_MU}{MENUBRANCH TRIO_MU}
9
10          /*GET ASSAY TYPE*/
11 GET_TYPE {MENUCALL TYPE_MU}
12          {IF FORM=0#AND#FUNC=0}{PUT FL_REC,1,1,TEST}
13          {IF FORM=0#AND#FUNC=1}{PUT BCH_FM,2,5,TEST}
14          {MK_TRIO}
15
16          /*GET RUN DATE*/
17 GET_DT   {GETLABEL "ENTER ASSAY DATE (MM/DD/YY): ",RUN_DT}
18          {MK_DT RUN_DT}
19          {IF @ISERR(BUFFERA)}{BEEP}{RETURN}
20          {IF FORM=0#AND#FUNC=0}{PUT FL_REC,2,1,BUFFERA}
21          {IF FORM=0#AND#FUNC=1}{PUT BCH_FM,2,6,BUFFERA}
22          {MK_TRIO}
23
24          /*GET BARCODE NUMBER*/
25 GET_BNUM {GETNUMBER "ENTER BATCH NUMBER: ",BATCH}
26          {IF FORM=0#AND#FUNC=0}{PUT FL_REC,3,1,BATCH}
27          {IF FORM=0#AND#FUNC=1}{PUT BCH_FM,6,5,BATCH}
28          {MK_TRIO}
29
```

Figure 9-15

the {GET_REC} and {KEY_IMP} subroutines. If you recall, the BCH_TRIO and RES_TRIO index files had index key expressions based on three fields: ASSAY_TYPE, RUN_DT, and BATCH. To locate records using these two index files, a concatenated string must be prepared. The string is prepared from data in the BCH_FM template. Directing the preparation is the TRIO_MU menu shown in Figure 9-15. The connection to the menu is made through the {MENUBRANCH} command in cell HA7 of Figure 9-13.

The TRIO_MU of Figure 9-15 allows current key values to be used or a change in any of the current key values. Depending on the key value being changed, TRIO_MU calls {GET_TYPE}, {GET_DT}, or {GET_BNUM}. These three subroutines are similar in function. The only difference is a customization of the type of data being input. If ASSAY_TYPE is input, the TYPE_MU from Figure 9-1 is called to allow user selection of the assay type. Using TYPE_MU prevents different users from specifying different names for test types. {GET_DT} uses a {GETLABEL} command to obtain a date string and then calls the {MK_DT} subroutine to convert the string to a Lotus serial date number. {GET_BNUM} uses a {GETNUMBER} command to prompt the user to input

```
--------EV---------EW---------------EX-----------------EY--------------------
29
30          /*BUILD KEY EXPRESSION FROM TRIO*/
31 MK_TRIO  {IF FORM=0#AND#FUNC=0}{TRIO_0}  /*USE FLUORESCENT DNA TEMPLATE*/
32          {IF FORM=0#AND#FUNC=1}{TRIO_1}  /*USE BATCH DETAIL TEMPLATE*/
33
34          /*PREPARE TRIO KEY FOR CHK_TRIO AND KEY_IMP FROM DATA*/
35                      /*IN FLUORESCENT DNA TEMPLATE*/
36 TRIO_0   {PAD "BATCH","ASSAY_TYPE",@INDEX(FL_REC,1,1)}
37          {LET KEY,BUFFERA}               /*INITIATE KEY*/
38          {LET BUFFERD,BUFFERA&"~"}  /*INIT COMMAND STRING FOR KEY_IMP*/
39          {PAD "BATCH","RUN_DT",@INDEX(FL_REC,2,1)}
40          {LET KEY,KEY&BUFFERA}           /*ADD DATE TO KEY*/
41          {LET BUFFERD,BUFFERD&BUFFERA&"~"}/*ADD DATE TO KEY_IMP STRING*/
42          {PAD "BATCH","BATCH",@INDEX(FL_REC,3,1)}
43          {LET KEY,KEY&BUFFERA}           /*ADD BATCH TO KEY*/
44          {LET BUFFERD,BUFFERD&BUFFERA}  /*ADD BATCH TO KEY_IMP STRING*/
45
46          /*PREPARE TRIO KEY FOR GET_REC AND KEY_IMP FROM DATA*/
47                      /*IN BATCH DETAIL TEMPLATE*/
48 TRIO_1   {PAD "BATCH","ASSAY_TYPE",@INDEX(BCH_FM,2,5)}
49          {LET KEY,BUFFERA}          /*INITIATE KEY*/
50          {LET BUFFERD,BUFFERA&"~"}  /*INIT COMMAND STRING FOR KEY_IMP*/
51          {PAD "BATCH","RUN_DT",@INDEX(BCH_FM,2,6)}
52          {LET KEY,KEY&BUFFERA}           /*ADD DATE TO KEY*/
53          {LET BUFFERD,BUFFERD&BUFFERA&"~"}/*ADD DATE TO KEY_IMP STRING*/
54          {PAD "BATCH","BATCH",@INDEX(BCH_FM,6,5)}
55          {LET KEY,KEY&BUFFERA}           /*ADD BATCH TO KEY*/
56          {LET BUFFERD,BUFFERD&BUFFERA}  /*ADD BATCH TO KEY_IMP STRING*/
57
```

Figure 9-16

a BATCH number.

Next, FORM and FUNC cells are tested and template-specific commands are issued to place the input data into the appropriate template. The {MK_TRIO} subroutine (Figure 9-16) is then called to make two strings. Depending on the values of FORM and FUNC, one of two subroutines are called ({TRIO_0} or {TRIO_1}). These subroutines create the following strings:

- A concatenated string to be used as a key value to be compared to the index key expression for the BCH_TRIO.NDX and RES_TRIO.NDX indexes. The final string consists of the current values to be used for ASSAY_TYPE, RUN_DT, and BATCH. Each sub-string is properly padded prior to concatenation. As the concatenated string is formed, it is placed into the KEY cell of the scratch-pad (Figure 9-21). When adding the next sub-string, the current value of KEY is retrieved from the cell, the concatenation occurs, and the new sub-string is placed back into KEY. The concatenation process is performed by the trio of

```
---------AV--------AW-------AX-------AY--------AZ--------BA-------BB---
 1  BCH_FM 0,0
 2  DATE:          ________                RECORD NUMBER:        ________
 3
 4                          ASSAY BATCH DETAILS
 5
 6  ASSAY TYPE:         __________         BATCH NUMBER:         ___
 7  DATE PERFORMED:     _________          TECHNICIAN:           _____
 8
 9  REAGENT LOT #:      _________
10  STANDARDS LOT #:    _________
11
12           *****AVERAGE FLUORESCENCE FOR STANDARDS*****
13  ________ _________          ________ _________
14  ________ _________          ________ _________
15  ________ _________          ________ _________
16
17           *****CURVE FITTING PARAMETERS*****
18  SLOPE:     ________         STANDARD ERROR Y ESTIMATE:   ________
19  INTERCEPT:________          R SQUARED:                   ________
20
21  COMMENTS:  ____________________         FINAL REVIEW DATE: ________
22
23                    *****SAMPLE DATA*****
24
25    Recno    BARCODE DIL_FCTR RAW_DATA1 RAW_DATA2  RESULTS PLATE_NUM
26  ________ ________ ________ ________ ________ ________ ________
27  ________ ________ ________ ________ ________ ________ ________
28  ________ ________ ________ ________ ________ ________ ________
29  ________ ________ ________ ________ ________ ________ ________
30  ________ ________ ________ ________ ________ ________ ________
31  ________ ________ ________ ________ ________ ________ ________
```

Figure 9-17

{LET} commands in cells EW37, EW40, and EW43 of {TRIO_0} and cells EW49, EW52, and EW55 of {TRIO_1}.

- A concatenated, padded, string containing program equivalents of carriage returns to be used as input for a call to {KEY_IMP}. If you recall from Chapter 8, when an index file contains an index key expression with multiple fields, @BASE prompts for each field in turn. After each field is typed, a [RETURN] is pressed and the program prompts for the next field name. {KEY_IMP} can be used to automate this process if {PAD} is used for each field's string and the strings are concatenated with "~" after all but the last string. This special concatenation is the function of the trio of {LET} commands acting on BUFFERD in cells EW38, EW41, and EW44 of {TRIO_0} and cells EW50, EW53, and EW56 of {TRIO_1}.

After the user is satisfied with the current search key values, the Current menu choice is

```
--------DN--------DO-------DP--------DQ--------DR-------DS-------DT--------
 1              /*PROGRAM TO EXPORT DATA FROM FLUORESCENT ASSAY*/
 2                       /*TEMPLATE TO BATCH DBASE*/
 3
 4   EXP_BCH    {HOME}{GOTO}BATCH_FM~         /*DISPLAY BATCH DETAIL FORM*/
 5              {WINDOWSOFF}{PANELOFF}        /*FREEZE THE DISPLAY*/
 6              {LET FORM,0}{LET FUNC,0}      /*SET FORM AND FUNCTION*/
 7              {MENUCALL TRIO_MU}/*GET ASSAY TYPE, RUN DATE, & BATCH NUMBER*/
 8              {SEL_NDX "BATCH","BCH_TRIO"} /*SELECT BATCH PRIMARY INDEX*/
 9              {CHK_TRIO}                    /*SEE IF RECORD EXISTS*/
10              {IF RECRD<>0}{MENUCALL OK_MU}/*RECORD EXISTS PROCEED?*/
11              {DB_EXP "BATCH",@UPPER("FL_REC")}       /*EXPORT FL_REC RANGE*/
12              {GET_BCH} /*DETERMINE RECORD NUMBER (MAY HAVE BEEN CREATED)*/
13              {DB_EXP "BATCH",@UPPER("FL_AVG")}       /*EXPORT FL_AVG RANGE*/
14              {UPDATE "BATCH",@UPPER("TECH_COMS"),TECH_COMS}
15              {FOR INC,0,6,1,EXP_STATS}     /*PUT STATS INTO RECORD*/
16              {EXP_SAMPS}          /*EXPORT SAMPLE DATA*/
17              {BRANCH STOP}        /*UN-FREEZE DISPLAY, RETURN TO FN_MU*/
18
19              /*SEE IF RECORD EXISTS WITH CONCATENATED TRIO AS THE KEY*/
20  CHK_TRIO    {GET_REC "BATCH",KEY}         /*SEE IF RECORD EXISTS*/
21              {PUT FL_REC,0,1,RECRD}        /*PUT RECORD NUMBER INTO TEMPLATE*/
22
23              /*PUT RECORD NUMBER INTO FL_AVG AND FL_CONC RANGES FOR EXPORT*/
24  GET_BCH     {CHK_TRIO}           /*GET RECRD: EXPORT MAY HAVE CREATED*/
25              {PUT FL_AVG,0,1,RECRD}        /*TRANSFER RECRD TO FL_AVG TEMPATE*/
26              {PUT FL_CONC,0,1,RECRD}       /*TRANSFER RECRD TO FL_CONC TEMPATE*/
27
28              /*TRANSFER STATISTICS DATA TO BATCH DATABASE*/
29  EXP_STATS   {MK_STATS}           /*ASSEMBLE THE STATS DATA FOR A FIELD*/
30              {UPDATE "BATCH",BUFFERD,@INDEX(STAT_TB,BUFFERB,BUFFERC)}
31
32              /*PLACE COLUMN, ROW, AND FIELD NAME INTO BUFFERS*/
33  MK_STATS    {LET BUFFERB,@INDEX(S_TAB,1,INC)} /*COLUMN COORD IN STATS AREA*/
34              {LET BUFFERC,@INDEX(S_TAB,2,INC)} /*ROW COORD IN STATS AREA*/
35              {LET BUFFERD,@INDEX(S_TAB,0,INC)} /*FIELD NAME TO UPDATE*/
36
37              /*PROGRAM TO EXPORT SAMPLE DATA FROM FLUORESCENT*/
38                       /*TEMPLATE TO RESULTS DATABASE*/
39  EXP_SAMPS   {WINDOWSOFF}{PANELOFF}        /*FREEZE THE DISPLAY*/
40              {GOTO}TARGET~        /*COPY ASSAY_TYPE,RUN_DT,& BATCH TO RECORDS*/
41              {BLANK TARGET}       /*REMOVE OLD DATA FROM TRIO AREA*/
42              /CFL_TRIO~.{DOWN @COUNT(BC_CT)-1}~
43              /RNCFL_REP~{ESC}.{RIGHT 8}{DOWN @COUNT(BC_CT)}~ /*REST RANGE*/
44              {DB_EXP "RESULTS",@UPPER("FL_REP")}
45              {HOME}
46              {BRANCH STOP}
```

Figure 9-18

selected from TRIO_MU and program control returns to {IMP_BCH}. {GET_REC} uses the concatenated string in KEY to determine the record number. After a record number is determined, all subsequent operations are the same as in previous example programs.

```
--------DN--------DO-------DP--------DQ--------DR-------DS-------DT--------
47                        STATISTICS TABLE OFFSETS
48                                (S_TAB)
49                       ==============================
50                       CURV_A               3         1
51                       CURV_B               2         7
52                       STD_Y                3         2
53                       R_SQR                3         3
54                       N_OBS                3         4
55                       DF                   2         5
56                       STD_C                2         8
```

Figure 9-19

The {KEY_IMP} subroutine uses the string with the tildes (" ~ ") to import tabular results into the template. As with previous example programs, {KEY_IMP} is preceded with an {IF} command to ensure that at least one record exists for the key value.

The {EXP_BCH} program in Figures 9-18 and 9-19 exports batch data from the template in Figure 9-20 to the BATCH.DBF database file. {EXP_BCH} is similar in form and function to {IMP_BCH} because it uses the TRIO_MU menu and MK_TRIO programs to assemble strings to be used for {GET_REC} and data transfer subroutines.

The {GET_REC} subroutine is called from {CHK_TRIO}. If the record exists, the record number is placed into cell A4. (The FL_REC range is A3..G4.) This placement is necessary for exporting data using {DB_EXP}. {DB_EXP} will overwrite existing records if the first field of the export range contains a non-zero Recno. If a record number is not found, a zero will be placed into cell A4 so that a new record will be created when {DB_EXP} is used.

Next, an {IF} command is used to test whether the record exists. If the record exists for a key value, the OK_MU menu from Chapter 8 is called to ask a user whether to proceed with overwriting the old record.

Then the {DB_EXP} subroutine is called. If you recall from Chapter 8, {DB_EXP} transfers records from a template to a database file based on key values in the spreadsheet. Selected fields matching a row of field labels in the template are imported. Because new records are created if Recno is in the field label range and the cell below Recno is zero, {DB_EXP} provides a very convenient method to create new records for database files indexed on multiple fields. Without {DB_EXP}, it would be difficult to create new records and place key values into all index fields necessary to look up the record. To illustrate this point, suppose {CHK_CRT} were used to create a new record. {CHK_CRT} could easily create a record and then use {UPDATE} to add a field of data to one of the keys. Next, the record number for the new record would need to be found before other key values could be added to the record. However, if the index had multiple fields in the index key expression, having two empty fields in key positions of the record would make a search of the index difficult and error-prone.

```
-------A--------B---------C---------D---------E--------F---------G----
 1                        FLUORESCENT DNA ASSAY DATA
 2  RECORD      TYPE     RUN DATE    BATCH      TECH   STANDARDS REAGENTS
 3   Recno   ASSAY_TYPE   RUN_DT     BATCH   RUN_TECH  LOT_STD   LOT_RGNT
 4     1436 DNA          11/27/90          3 LMM       24XLM03   55D1090
 5 TECH_COMS 24-HOUR INCUBATION
 6
 7                        ****STANDARD CURVE DATA****              X
 8 NG DNA/ML   ASSAY1     ASSAY2     ASSAY3    ASSAY4    AVERAGE   PREDICT
 9         0      93.9       92.9       94.3      94.3      93.9      -0.4
10        25     130.4      134.6      133.4     135.9     133.6      26.1
11        50     174.5      171.8      173.1     174.1     173.4      52.8
12        75     202.9      204.5      204.2     205.2     204.2      73.4
13       100     235.0      242.2      242.7     239.9     240.0      97.3
14       250     473.9      462.0      469.3     473.2     469.6     250.9
15
16                     ****CONCENTRATIONS OF STANDARDS****
17 Recno        CONC_0    CONC_1     CONC_2    CONC_3    CONC_4    CONC_5
18     1436          0        25         50        75       100       250
19
20                        ****AVERAGE FLUORESCENCE****
21 Recno        AVG_S0    AVG_S1     AVG_S2    AVG_S3    AVG_S4    AVG_S5
22     1436       93.9     133.6      173.4     204.2     240.0     469.6
23
24                ==============================================
25                |           Regression Output:               |
26                |  Constant                         94.518421|
27                |  Std Err of Y Est                 3.3328948|
28                |  R Squared                        0.9994979|
29                |  No. of Observations                      6|
30                |  Degrees of Freedom                       4|
31                |                                            |
32                |  X Coefficient(s)     1.494878             |
33                |  Std Err of Coef.     0.016751             |
34                ==============================================
35
36             *********SAMPLE DATA SECTION***********
37
38             BARCODE  DIL_FCTR RAW_DATA1 RAW_DATA2  RESULTS PLATE_NUM
39             2041207        40     200.3     209.7     2956    273460
40             8268367        10     368.1     377.7     1862    273460
41             7395883         1     290.0     303.3      135    273460
42             7985840        20     427.9     428.4     4464    273460
43             1571806        10     408.9     423.2     2151    273460
44             1849327       100     106.8     113.6     1049    273460
45             7779731        40     161.8     169.1     1898    273460
46             4647388       100      94.9     102.1      266    757499
47             1250091         1     281.0     290.5      128    757499
48             7793121        20     351.8     362.4     3513    757499
49             4776316        40     250.8     269.1     4427    757499
50             5753781        40     173.5     175.7     2143    429393
```

Figure 9-20

After the values in FL_REC are exported, the index file is searched again to determine the record number. This search (performed in the {GET_BCH} subroutine) is necessary because {DB_EXP} may have created a new record. This second search will find the record number of the new (or pre-existing) record. Next, a pair of {PUT} commands in {GET_BCH} transfer the record number to the FL_AVG and FL_CONC ranges. Placement of the record number into these ranges is necessary so that the data in the ranges will be added to the relevant record when {DB_EXP} is used. Without these record numbers, new records would be created.

The {EXP_SAMPS} subroutine in Figure 9-18 illustrates some other new concepts. As was stated in Chapter 8, {DB_EXP} requires a range to export. The range must start at the row containing the field names and extend down the data to be exported. The range cannot be open-ended. It the range is larger than the table of data to be exported, empty records are created. If the range is smaller than the table of data, not all of the records will be exported. The /RNC command in cell DO43 re-creates the FL_REP range to meet this rule. The @COUNT(BC_CT) argument in the {DOWN} command counts the number of barcodes in the barcode count range (BC_CT, B39..B999) and supplies the number to the {DOWN} command. When the command in cell DO43 executes, the {DOWN} command highlights the appropriate number of cells. It would be good practice to manually issue the commands specified in cell DO43 to get a better understanding of each part of the commands.

Another important command sequence is in cell DO42. This sequence copies the ASSAY_TYPE, RUN_DT, and BATCH information from the FL_TRIO range (B4..D4) to each record in the Sample Data section. If you recall, this trio is needed for one of the two index files associated with the RESULTS.DBF database (i.e., RES_TRIO.NDX). By copying the FL_TRIO to each record, the program sets up records for future retrieval into a batch detail form using relational techniques.

{DB_EXP} in {EXP_SAMPS} exports sample result data (including the FL_TRIO data) to the RESULTS.DBF database. As suggested above, it would be good practice to manually issue the commands in cell DO42. Before issuing the commands, use the {GOTO} key ([F5]) to go to the TARGET cell (Cell H39).

THE SCRATCH-PAD

The scratch-pad for the above programs is shown in Figure 9-21. As explained in Chapter 6, a scratch-pad contains a set of cells that temporarily store data. Each of these cells is called a buffer. A buffer is merely a temporary storage area for data. A buffer is usually empty or contains erroneous data at the beginning of a program. As programs execute, data need to be temporarily stored, and the data are stored in buffers. When more data need to be stored, the new data overwrite the current contents of the buffer cells. The following are brief descriptions of each buffer in the scratch-pad.

- FORM is a buffer cell that holds data from the menu hierarchy. The information stored in FORM is: specifications for the type of data to manipulate, the template, database, and index files to be used, etc.

```
--------EK--------EL--------EM-------EN-------EO-------EP-------EQ-------ER---
 1
 2                         /*SCRATCH-PAD CELLS*/
 3
 4  FORM                 0 /*TYPE OF DATA TO MANIPULATE*/
 5  FUNC                 0 /*FUNCTION: 0=EXPORT;1=IMPORT;2=EDIT;3=PRINT*/
 6  SUB_PTR  RGT_EXP       /*SUBROUTINE POINTER CELL FOR {DISPATCH} COMMANDS*/
 7
 8  TEST     DNA           /*ASSAY TYPE*/
 9  RUN_DT   11/27/90      /*RUN DATE*/
10  BATCH                3 /*BATCH NUMBER*/
11  KEY      DNA       33204  3  /*INDEX KEY TO USE FOR SEARCH*/
12
13  INC                 21 /*LOOP COUNTER*/
14  TRAP                 1 /*TRAP TO RECEIVE ERROR CODE FROM @DB AND @NDX FUNCS*/
15  RECRD                7 /*RECORD NUMBER*/
16  BCODE              101 /*BARCODE NUMBER*/
17  BUFFERA  0.01675195 |*GENERAL PURPOSE BUFFERS:       *|
18  BUFFERB          2 |*BUFFERA IS THE ONE THAT IS USED*|
19  BUFFERC          8 |*BY MOST SUBROUTINES TO RETURN  *|
20  BUFFERD  STD_C      |*VALUES TO THE CALLING FUNCTIONS*|
21  FLD_NM   STD_C         /*FIELD NAME FOR @DB FUNCTIONS*/
22  DB_FILE  BATCH         /*DATABASE FILE TO USE*/
23  NDX      BCH_TRIO      /*INDEX FILE TO USE*/
24
```

Figure 9-21

- FUNC holds data generated from the FN_MU and codes for the type of function to be performed. More specifically, values of 0, 1, 2, and 3 will appear in FUNC for input, review, editing, and printing of records, respectively.
- SUB_PTR is an acronym for "subroutine pointer". SUB_PTR holds a character string for the name of the subroutine that a {DISPATCH} command will branch to when the command executes.
- TEST is a buffer that holds data corresponding to the ASSAY_TYPE field. The buffer receives its value from the TYPE_MU and its associated submenu system.
- RUN_DT is a buffer that contains the assay date.
- BATCH contains the batch number being used.
- The KEY cell is a buffer that holds index key values used for index file searches. KEY also holds intermediate strings when strings are being concatenated to prepare index key values containing multiple field names.
- INC is a buffer used as a loop counter in {FOR} commands. INC often concurrently provides offset values for @INDEX @functions and {PUT} commands.
- The most convenient way to use @DB and @NDX @functions in macro programs is to place them into {LET} commands. However, {LET} commands require cell locations to place values into. TRAP receives data from the {LET}

commands, thereby satisfying this requirement.

- RECRD is a cell that holds the current record number.
- BCODE is a cell that receives barcode numbers from {GETNUMBER} commands and other program functions.
- BUFFERA through BUFFERD are general-purpose buffers and are used as the electronic equivalents of electronic scratch-pads. Of the buffers, BUFFERA is perhaps the most important because most of the Generic Subroutine Utilities described in Chapter 8 return their output to BUFFERA. Except for this special purpose, any of the buffers can be used for short-term storage of any data necessary, as long as you keep track of which data are in which buffer so that your program uses the correct information. To this end, the following loosely defined guidelines are used: Buffers B and C are usually used for column and row offset values for @INDEX and {PUT} commands. BUFFERD is often used for field names and strings for {KEY_IMP}.
- FLD_NM is a cell that often holds the field name for @DB and @NDX @functions.
- DB_FILE is a cell that holds the name of the current database file being accessed.
- NDX is a cell that holds the name of the current primary index file being used.

CREATING THE EXAMPLE PROGRAMS

Create the programs in Figures 9-2, 9-3, 9-5, 9-6, 9-8, 9-10, 9-11, 9-13 through 9-16, 9-18, 9-19 and 9-21 by entering the text into your spreadsheet. To activate the programs, issue the /***R***ange ***N***ame ***L***abel ***R***ight command and highlight the cells containing the subroutine names in Column HL. Press [RETURN]. Repeat for cells containing macro names in columns DA, FR, GO, GD, GZ, EV, DN, and EK.

The report form templates and ranges for the report forms were created in Chapter 2. However, named ranges for the tables containing column and row offsets for the forms still need to be created. Create the following ranges using the /***R***ange ***N***ame ***C***reate command:

Range Name	Range of Cells
C_TAB	DC28..DE33
STD_TAB	FS63..FU84
SI_TAB	GQ41..GS52
SRD_TAB	GF42..GH63
BD_TAB	HB32..HD55
S_TAB	DP50..DR56

When designing your application, map out the column and row offsets for each piece of data in your report form templates. Start by creating a range name for each template. Place a form name and "0,0" in the cell at the top-left corner of each report form range. This cell represents the origin from which all column and row offsets are measured. Print

```
----------CQ--------CR--------CS----------CT-------CU---------CV----------CW---
 1
 2
 3                          RANGE NAME TABLE
 4
 5          ABORT      FH42                     FN_MU      CH47
 6          ASMBL_STD  FS53                     FORM       EL4
 7          AUTOEXEC   BV25                     FUNC       EL5
 8          BATCH      EL10                     FUNC_TAB   HO35..HO41
 9          BCH_FM     AV1..BB22                GET_BCH    DO24
10          BCH_HDR    AV25..BB25               GET_BNUM   EW25
11          BCH_TAB    AV26..BB106              GET_DT     EW17
12          BCODE      EL16                     GET_FLD    FH70
13          BC_CT      B39..B999                GET_NEXT   FH74
14          BD_TAB     HB32..HD55               GET_PREV   FH78
15          BUFFERA    EL17                     GET_REC    FH57
16          BUFFERB    EL18                     GET_SID    GP35
17          BUFFERC    EL19                     GET_TYPE   EW11
18          BUFFERD    EL20                     IL_MU      CH34
19          B_RNF      HA26                     IMP_BCH    HA4
20          B_SD       AV25..BB25               IMP_FILE   FH85
21          CALL_BASE  FH29                     IMP_KEY    FH89
22          CERT_FM    Y1..AF15                 IMP_LBLS   FH87
23          CERT_HDR   Y19..AF19                IMP_SRD    GE36
24          CERT_TAB   Y20..AF30                INC        EL13
25          CHK_CRT    FH48                     INTERCEPT  F26
26          CHK_TRIO   DO20                     KEY        EL11
27          C_RNF      DB23                     KEY_EXP    FH104
28          C_TAB      DC28..DE33               KEY_FILE   FH106
29          DB_EXP     FH94                     KEY_IMP    FH83
30          DB_FILE    EL22                     KEY_RNG    FH108
31          DB_MU      CH9                      LM_MU      CH30
32          DNA_TRIO   H39..H50                 MAIN_MU    CH3
33          DO_CERT    DB3                      MK_DT      FH17
34          DO_FUNC    HM6                      MK_STATS   DO33
35          DO_SRD     GE4                      MK_STRNG   FH12
36          DO_STD     FS6                      MK_TRIO    EW31
37          DO_S_ID    GP5                      NDX        EL23
38          EXP_BCH    DO4                      NEXT_SRD   GE20
39          EXP_FILE   FH96                     NPD_MU     HZ9
40          EXP_RNG    FH98                     O          BV14
41          EXP_SAMPS  DO39                     OK_MU      HZ4
42          EXP_SID    GP19                     PAD        FH22
43          EXP_STATS  DO29                     PDGF_MU    CH26
44          FIN_RGT    FH7                      PREV_SRD   GE25
45          FLD_NM     EL21                     PRINT      HM11
46          FL_AVG     A21..G22                 PRN_SPEC   HM16
47          FL_CONC    A17..G18                 PRN_TAB    HO24..HO30
48          FL_REC     A3..G4                   PUT_BCH    HA22
49          FL_REP     B38..J50                 PUT_SAMP   DB19
50          FL_TRIO    B4..D4                   PUT_SID    GP29
```

Figure 9-22

```
----------CQ--------CR--------CS----------CT-------CU---------CV---------CW---
52
53                            RANGE NAME TABLE
54
55            PUT_SRD     GE30                      STD_EXP     FS24
56            PUT_STD     FS38                      STD_FM      BF1..BP20
57            RAW_STD     B9..E14                   STD_TAB     FS63..FU84
58            RECRD       EL15                      STOP        FH38
59            RELEASE     BV48                      SUB_PTR     EL6
60            RET_BASE    FH31                      S_TAB       DP50..DR56
61            RGT_LP      FH6                       TARGET      H39..J499
62            RG_FM       BF25..BP45                TECH_COMS   B5
63            RG_MU       CH40                      TEMPLATE    A1..I51
64            RUN_DT      EL9                       TEST        EL8
65            SEL_NDX     FH34                      TRAP        EL14
66            SID_FM      O1..U19                   TRIO_0      EW36
67            SI_TAB      GQ41..GS52                TRIO_1      EW48
68            SLOPE       E32                       TRIO_MU     EW5
69            SND_STDA    FS44                      TST_CELL    BV47
70            SND_STDB    FS48                      TYPE_MU     CH22
71            SND_STDC    FS58                      UPDATE      FH65
72            SRD_FM      AK1..AR25                 USR_INP     FH5
73            SRD_NF      GE17                      VU2_OPEN    BV44
74            SRD_TAB     GF42..GH63                VU_OPEN     BV40
75            SR_MU       CH14                      \0          BV8
76            STATS       C25                       \M          HM44
77            STAT_TB     C25..F33
```

Figure 9-23

the forms. Then count the column and row offsets of each cell that will receive field data from the macro program interacting with the form. Using these offsets and database field names, create an offset table for each form. Example tables are shown in Figures 9-14 and 9-19. Create named ranges for each offset table using the /***R***ange ***N***ame ***C***reate command. The range to specify is the one starting at the first field name and ending with the last field name.

Create the following ranges using the /***R***ange ***N***ame ***C***reate command:

Range Name	Range of Cells
PRN_TAB	HO24..HO30
FUNC_TAB	HO35..HO41

Figures 9-22 and 9-23 show a printout of the Range Name table for the forms, tables, macro programs, menus, scratch-pad buffers, and generic subroutine utilities created in Chapters 3, 7, 8, and 9. Issue the /***R***ange ***N***ame ***T***able command to create a table in your spreadsheet and compare it to Figures 9-22 and 9-23 to ensure that all ranges have been created and defined correctly. If there are any discrepancies, resolve them. When finished, delete the columns containing the Range Name table. The Range Name table

```
--------IL-------IM-------IN-------IO-------IP-------IQ-------IR-------IS---
 1
 2            /*PROGRAM TO ILLUSTRATE USE OF CRITERIA*/
 3
 4  EOM       {GETLABEL "PLEASE ENTER BEGINNING DATE: ",BUFFERA}
 5            {MK_DT BUFFERA}   /*CONVERT STRING TO SERIAL DATE NUMBER*/
 6            {MK_STRNG BUFFERA}/*CONVERT DATE NUMBER TO STRING*/
 7            {LET BUFFERD,+"RUN_DT>"&BUFFERA&"#AND#RUN_DT<"}
 8            {GETLABEL "PLEASE ENTER ENDING DATE: ",BUFFERA}
 9            {MK_DT BUFFERA}   /*CONVERT STRING TO SERIAL DATE NUMBER*/
10            {MK_STRNG BUFFERA}/*CONVERT DATE NUMBER TO STRING*/
11            {LET BUFFERD,BUFFERD&BUFFERA}
12            {DB_COUNT "RESULTS",BUFFERD}
13                      •
                        •
                        •

--------FG-------FH-------FI-------FJ-------FK-------FL-------FM-------FN---
115
116
117           /*OBTAIN NUMBER OF RECORDS MEETING CRITERIA*/
118 DB_COUNT {DEFINE DB_FILE,BUFFERB:VALUE}
119           {LET BUFFERA,@DBCNT(DB_FILE,BUFFERB)}
120
```

Figure 9-24

is large and consumes a lot of memory. Deleting the columns releases the memory. Next, insert the equivalent number of columns so that the spreadsheet again matches the example in this book.

For practice, design and create a macro program to transfer data between the RG_FM (Reagents Form) of Figure 3-9 and the REAGENTS database. The program should be similar to the program for STD_FM and the STANDARD database and should be called DO_RGT. The pattern program for DO_RGT is shown in Figures 9-5 and 9-6.

USING CRITERIA

For more practice, create a program to generate end-of-month reports. Base the program on the {DB_COUNT} program you designed in Chapter 8. Figure 9-24 illustrates a portion of the program that you need to design. The following are some details that should be of help to you:

- The beginning and ending dates are obtained from the user with {GETLABEL} commands in cells IM4 and IM8, respectively.
- The dates are converted to Lotus serial date numbers using {MK_DT}. (If you recall, the RUN_DT field was defined as numeric in the RESULTS.DBF

database.)

- Lotus serial date numbers are converted to strings using {MK_STRNG}. Conversion to strings is necessary because the date numbers will be concatenated into a string containing field names and the #AND# operator. If numbers were used, an error would occur.
- The string concatenation is performed by the {LET} commands in cells IM7 and IM11. To illustrate the result of the concatenation, if 12/1/90 and 12/31/90 were specified, the following string would be generated in BUFFERD:

```
RUN_DT>33208#AND#RUN_DT<33238
```

- Note the plus sign (+) in the {LET} command of cell IM7. This sign is of critical importance. When using {LET} commands to concatenate, you must precede the equation with a plus sign (+) if strings within quote signs are being used. When only spreadsheet cell names are being used, you do not need to use the plus sign, as illustrated in the {LET} command in cell IM11.
- The {DB_COUNT} subroutine is called to determine the number of records meeting the criteria in BUFFERD.
- An example DB_COUNT program is illustrated in Figure 9-24.

Create a report form and finish the EOM program to ensure that you understand the concepts of this chapter and previous chapters. Another good exercise would be to include a third criterion in the expression (e.g., R_STATUS).

SUMMARY OF WHAT HAS BEEN ACCOMPLISHED

This chapter showed how to design and create programs that coordinated the use of the subroutine building blocks created in Chapter 8 to form a system that moved data between database files and spreadsheet templates. The following are some important highlights of the example programs:

- An emphasis was placed on creating organized macro programs. Each program will normally have the same (or similar) subsections. You should try to align subsections in the same order. Aligned subsections lead to more error-free programs because the chances of forgetting important programming steps is minimized.
- Examples of how to pass correctly formatted arguments to generic subroutine utilities were provided.
- Tables were used to provide column and row offsets for {PUT} commands and @INDEX @functions. The tables also contained field names for efficient transfer of data between databases and templates.
- Primary index files were selected near the beginning of all macro programs. Although this selection may not be necessary in all instances, the action should be taken anyway. By specifically stating the primary index files, programs will

be easier to interpret and will be independent of other, previously executed, programs.

- {PAD} was used to pad spaces into strings. {PAD} was ***NOT*** used for fields specified as numeric and not part of index key expressions containing multiple field names.
- {IF} commands were used prior to calling generic subroutine utilities based on @BASE menu commands. These {IF} commands were necessary to eliminate program termination when no records were found for the key value.
- {LET} commands were positioned before {DISPATCH} commands and were used to specify the macro program to branch to when the {DISPATCH} command was executed.
- When using {LET} commands to concatenate, you must precede the equation with a plus sign (+) if strings within quote signs are being used.

WHAT'S NEXT?

The next chapter will give you a review of what you have learned in previous chapters and will give you some pointers on designing databases, building templates, programming, troubleshooting problems, etc.

10

Programming Tips and Troubleshooting Guide

Trouble-free and efficient databases, templates, and macro programs require conscientious designs. Some of the tips that I will give you have been mentioned in previous chapters; others are new. Bringing these tips together in one section will give you a one-stop checklist of topics to consider as you are creating your databases, templates, and macro programs.

SYSTEM CONFIGURATION

Before designing and creating your LIM system, you should be aware of memory and open file limitations in your computer system. The following two subsections provide reviews of the most important of these limitations.

Memory Considerations

Lotus, @BASE, and your spreadsheets are all stored in your computer's conventional memory. The @BASE data files are stored on disk. When working with large data files, storing them on disk rather than in the memory allows you to store many more data records.

However, @BASE and the @BASE Option Pac add-in application programs occupy substantial amounts of conventional memory, limiting the size of the spreadsheet you can create before you obtain a "Memory Full" message. Therefore, it is important for you to determine the amount of available memory after all of your application programs have been added in, verify that enough conventional memory remains for your spreadsheet, and modify your system accordingly. Modification suggestions were provided in Chapters 1 and 4.

Additionally, if you are using Symphony 2.2, you should check to ensure that you have not added both BASEFUNC.APP and BASEOFTF.APP. If BASE-FUNC.APP has been inadvertently added, remove it. BASEFUNC.APP is not needed and consumes memory. For details on how to remove BASEFUNC.APP, see Chapter 4.

If, after all of the above suggestions have been implemented, your application is still low on memory, enable your EMS memory.

Open File Considerations

When designing and creating a database and index system, there is a very important DOS limitation that you need to be aware of. Although DOS allows up to 255 files to be opened, only 20 files can be accessed by each application. Because @BASE runs within Lotus, the two programs together cannot access more than 20 files, regardless of how many files are specified in your CONFIG.SYS file. Lotus and @BASE typically have about 6 open files, leaving about 14 user files. ***Thus, you can have a total of up to about 14 open database and index files before the system begins to issue error messages.*** As configured in this book, the example system has five database files (REAGENTS, STANDARD, BATCH, SAMPLES, and RESULTS) and six index files (SAMPL_BC, RES_BC, RES_TRIO, BCH_TRIO, STDS, and RGNTS). It is important to allow at least one unused file so that the spreadsheet can be saved. Therefore, only $(14-5-6-1) = 2$ more open database or index files would normally be available.

CONNECTIVITY: SUPPORT FOR LOCAL AREA NETWORKS

The database and index files created by @BASE are fully compatible with the dBASE® III and dBASE III Plus file formats. Because dBASE III is Local Area Network (LAN)-compatible, your system can easily tie into LANs using dBASE III.

Symphony 2.2 contains features that enhance its usage in LANs. These features are incorporated into the ways that files are opened. Release 2.2 allows you to open database files on disk according to type of access, a feature required by LANs. The type of processing you can perform depends on the type of access you have to database files. The following are the types of access allowable:

- ***E***xclusive: Opens the specified database file for reading and writing data and for performing commands that change the structure of the file. Selecting ***E***xclusive prevents other users on the LAN from opening the file until you close the file. To use ***F***ile ***F***ield-rename, ***F***ile ***M***odify, ***D***ata ***D***elete ***P***ack, ***D***ata ***D***elete ***D***elete-all, and ***D***ata ***S***ort ***O***verwrite, you must select ***E***xclusive access.
- ***P***rimary: Opens the specified database file for reading and writing data, but not for performing commands that change the structure of the file. Other users on the LAN can open the file with read-only access until you close the file. To use ***D***ata ***D***elete ***M***ark, ***D***ata ***D***elete ***U***nmark, ***D***ata ***R***eplace, ***D***ata ***S***ort ***C***reate-file,

***D**ata **T**ransfer **C**opy, **D**ata **T**ransfer **E**xport and **U**tility **F**ile **I**mport, you must select either **E**xclusive or **P**rimary access. You must also select either **E**xclusive or **P**rimary access to perform the following tasks:

- Editing or adding records to a file using **D**ata **B**rowse or **D**ata **F**orm.
- Marking or unmarking records for deletion using **D**ata **B**rowse or **D**ata **F**orm.
- Using @DBAPP, @DBDEL, and @DBUPD in a macro program.

- **R**ead-only: Opens the specified database file for reading only. You can view the file, but cannot change the data in, or structure of, the file.

If you select **E**xclusive or **P**rimary and another user has already opened the file with one of those options, @BASE will display an error message and you will be allowed to open the file with **R**ead-only until the other user closes the file.

DATABASE AND INDEX FILE DESIGN TIPS

Database definitions can be modified at any time. However, changes are risky and it is easy to lose data as a result of changes. Therefore, it is very important to plan your system prior to implementation. It is also important to prepare backup copies of your database files prior to changing definitions. That way, if data are lost, you can always go back to previous versions and retrieve the lost data.

So, whether you are creating a database to simply archive data from a single laboratory or assay or are creating a fully functional LIM system, take time to plan. The following are some guidelines to follow when designing your database system:

Number of Open Database and Index Files

When deciding on databases, observe the DOS open file limitation described above. ***You can have a total of up to about 14 open database and index files before the system begins to issue error messages.*** If your system exceeds the 14-file limit, you could decrease the number of files by two more if you merge databases containing similar records. However, merging files will result in slightly more difficult programming.

Field Names

Perhaps the most critical step in defining databases is that you need enough fields to ensure that each record in the database is unique. Redundancy in databases causes failures during access and cross-reference of information. Therefore, you need to ensure that each database has one or more fields that, combined, give unambiguous uniqueness to each record.

After you have decided on field names to eliminate record redundancy, you need to use your first level lists to define field names for the remainder of the information you want to keep track of and also the field names that you want to use for joining databases.

The apportionment of Information Fields between database files is an iterative process. Field names often need to be shuffled from one file to the next before you find the right combination of pigeon holes to achieve optimal system performance and reliability.

Indexing and joining database files often reveal redundancies and inconsistencies in field names and types. These problems are especially apparent when joining files. Although two or more database files can use identical field names, if the field names become virtual fields of the same file, @BASE will issue an error message. For this reason, it is good policy to review the lists of fields for each database file to ensure that they are correct and make modifications if needed.

The following rules apply to field names:

- Field names within a file ***must be unique***.
- Field names used to create links between two files when they are joined ***must be identical*** in both files.
- Names of lookup fields selected for viewing as a result of a join ***must be different*** than all of the field names existing in the primary file obtaining the information and any lookup fields from any other file joined to the primary file.
- Field names in macro programs, database files, and index files that function as specifications for data to be transferred between databases and spreadsheet templates or other databases, ***must be identical***. For example, ASSAYTYPE and ASSAY_TYPE would be treated as two completely different entities by both Lotus and @BASE. If a macro program tried to access ASSAY_TYPE in a database file by calling it ASSAYTYPE, an error message would be issued.

To ease compliance with the above rules, organize your fields into two categories and assign names according to category:

- ***Identification Fields:*** Fields that will be used as a basis for searching, sorting, and joining databases. The Identification Fields section can be thought of as the bookkeeping portion of a database file.
- ***Information Fields:*** Fields that will be used for archiving data for later retrieval. The Information Fields section can be thought of as the record-keeping or data archiving portion of a database file.

As you create the lists of fields in the databases for your system, it is very important to continually and thoroughly review all of the lists. When you are finished, review the lists together as a set. You should look for

- inconsistencies in field names.
- repeating fields within database records.

- repeating fields between database files.
- mixed relations within database records.
- redundancy in fields within database records.
- redundancy in fields between database files.
- inconsistencies in the flow of information between joined databases.
- situations that could cause records to lose their uniqueness.
- fields that are unlikely to be used.

Avoid Using Date Field Types in Indexes

There is a character limit that you cannot exceed when creating index files. The number of characters in equations for indexing depends on the number of fields being joined, the field types, and the length of the field names. Date fields consume a large number of characters before they can be used in indexing equations. Therefore, ***specifying dates for fields to be used in index files should be avoided.***

There is another reason to avoid dates in indexes. It is important to keep index equations as small and compact as possible. An index equation is re-evaluated for each record in the index file. If there are thousands of records in a file, a substantial amount of time (and conventional memory) may be required. Dates require six functions to be executed before dates are converted to usable form. If you build your own user interface, converting ***dates used as keys*** to numbers with the Lotus @DATE @function prior to storage is highly recommended. If you decide to use the @BASE or dBASE III user interface, the choice is a little more complicated. Dates will not be displayed in date format unless the field type is specified to be date. Therefore, user input and review would be impaired because the user would have to work with number equivalents of dates.

Field Type

Choosing the correct field type is very important. If you change a field type later, the existing data in the field will be converted to the new field type. If data types are incompatible, the data for a field will be lost. (For conversion tables, see Chapter 4.)

It is also critically important to have field types consistent between files that are to be joined. For example, if two files are joined using lot numbers defined as character in one database file and numeric in another database file, the information link would not be effective.

Index Files Used for Joining Databases

You will need one index file for each anticipated join application. The following approach to generating the list is one that works quite efficiently:

- Orient yourself in the correct direction. One lookup file record can service several primary file records. The lookup file contains unique information.
- Determine the primary and lookup database files you want to join from a flow

diagram.

- Review the Identification Fields sections of the lookup files and determine the appropriate field names to use for the joining process.
- Check the corresponding primary files to ensure that they have the exact same field names, types, and widths for the fields to be used for joining the two databases.
- If the join field names do not appear in the Identification Field sections of ***both*** files to be joined, see if the fields exist, but have been assigned different names. If named differently, resolve the problem by renaming the field in one of the files. If the fields do not exist in the Identification Fields section, look for the field names in the Information Fields sections. If the field names do not exist there, add a new field to the Information Fields section to provide join capabilities.

To reiterate the most important items on the list: It is important that you do not forget that you can only get one record from a lookup file and that field names, types, and widths used to join files must be identical.

TEMPLATE DESIGN TIPS

Plan your spreadsheet application before typing it in your template. Remember, the template not only needs to serve as a report, but must also work in harmony with the macro that runs it. Plan ahead. There is no substitute for sufficient planning. The time that you spend planning your template will be more than offset by the time you will save getting it to run correctly. If you do not plan your spreadsheet, you will probably have to make changes that may require inserting rows, deleting columns, moving data to different cells, etc. Re-designing templates and rewriting macro programs take a lot of time and lead to errors. To say the least, a well-planned template will lead to a more efficient macro program.

The following is a list of some review and supplementary style guidelines to consider when designing templates to be used with databases:

- Always design template sections so that they are formatted with efficient database transfer in mind. Transfer of data between spreadsheets and database files is usually the slowest process in a LIM system. Unfortunately, transfer of data is also one of the functions that is most often performed.
- When designing templates, do so while looking at the lists and data flow diagrams created in Chapter 2 to ensure that all appropriate fields of data are included in the template.
- Whenever possible, format data in tables that conform to the rules imposed by @BASE for menu transfer of data. Menu transfer of data is faster and often easier than @DB and @NDX @function transfer because of inefficient Lotus {PUT} commands and @INDEX @functions.

- A table used for @BASE menu transfer of data should consist of a header of field names (with the first field name being "Recno" if transfer is record number-based) followed immediately by a body of data. Further, there must be no blank rows or columns within the table.
- When you design your templates, determine the location for all data. From this design, determine the locations for the macro programs. This part of your plan is important. You do not want data overwriting program cells because overwriting that will destroy the program. You will also want to take the time to organize the subsections of the template so that the macro can efficiently place data into them.

  Organizing templates is especially important when importing tables of data from databases. Importing tables of data overwrites everything in the cells that fall within the output range. So, be sure to leave enough space below the row of import definition labels to account for the number of records being imported.
- When using field names across the top of a table for @BASE menu transfer of data, all field names need not appear in the table or be in the same order as the database, but all field names ***must*** be spelled identically to those in the database. Because field names tend to be acronyms and therefore somewhat confusing, it is often desirable to hide the field names and use more descriptive names in the row immediately above the field name row. For Lotus 1-2-3, the command sequence for hiding a range is /***R***ange ***F***ormat ***H***idden. For Symphony, the command sequence is /***F***ormat ***O***ther ***H***idden. (This treatment has not be applied to the templates in this book so that you can see the field names. However, under normal circumstances range names would be hidden.)
- If virtual field names are used, the names must be in lower-case letters.
- In templates are involved in record number-based transfer of data, do not forget to include fields in the spreadsheet for a Recno.
- The more formulae that you have in cells, the slower the spreadsheet will recalculate. If you have several hundred formulae in a template, every time the spreadsheet recalculates it will take a substantial amount of time. Therefore, if you need to place large data summaries in a spreadsheet, you should have a program do as many of the calculations as is practical.
- It is especially important to consider the above guideline when using @BASE @DB @functions. Each time Lotus performs a spreadsheet recalculation, @BASE recalculates all @DB @functions in the spreadsheet. Every time an @DB @function is evaluated, @BASE starts at the beginning of the database file and reads each record until it locates the information it is searching for. This process is typically very slow. For example, it would take approximately 9 seconds to evaluate a search on 1,000 records in a database on a standard IBM AT running at 6 Mz. If a Lotus spreadsheet contained many @DB @functions, and if a database file contained thousands of records, the spreadsheet would take a very long time to recalculate. ***Therefore, you should avoid at all costs the use of @DB @functions in spreadsheet cells.*** Instead, "nest" the functions in {LET} macro commands. By doing so, the @BASE @functions will be

calculated only when macro commands execute and spreadsheet recalculation will not be affected.

- @BASE @NDX @functions are much faster than @DB functions. For reasons explained in Chapter 2, @NDX @functions are nearly instantaneous. Therefore, @NDX use in cells does not significantly slow spreadsheet recalculation. However, @NDX @functions are more commonly nested in {LET} commands of macro programs, rather than used directly in cells of spreadsheets.
- If long lists of records need identical information in a table used to export data, you should first ask whether the data could better be contained in a joined database. If the information could not, then the data should be copied to columns immediately to the right of the existing table. The data could then be concealed by hiding the columns.
- If data are to be used by the Lotus regression program, the data must be in columns, not rows. This placement is at odds with @BASE menu-based transfer of data, because data must be in rows not columns for this type of transfer. Therefore, a compromise must be made. Either the transfer must be made using @DB and @NDX @functions from a column, or the template needs to contain the data in both formats. Having both formats is redundant, but makes database transfer faster.
- If a date is needed in a template, the date can be obtained from the Lotus @NOW @function. If the @function is placed in a cell of the template, it will update continually. If the @function is placed in a macro program, it will update only when the macro program executes. Therefore, you must decide how you want to use the date. If the current date is all that is needed, then it can be placed in the template cell. If the date is to be used for database update, it should be placed in a macro program.
- If a cell will be displaying a date, the cell must have a width of at least 9 characters and be formatted correctly. To format dates in Lotus 1-2-3, use the /***R***ange ***F***ormat ***D***ate ***4***(Long Int'l) command. To format dates in Symphony, use the /***F***ormat ***D***ate ***4***(Full Int'l) command.

The following subsections outline some other topics to consider when designing templates. These considerations apply to template design whether they are used with databases or not.

Allow for Expansion

As projects proceed, the amount of data that is collected in each experiment usually increases as new ideas are generated. This expansion is a way of life in most laboratories. If you plan your template for expansion from the start, you can reserve room on the spreadsheet for extra rows and columns that can be used to hold extra data.

Bigger Is Not Necessarily Better

Placing too many different types of experiments in one spreadsheet can make it cumbersome, thus making it confusing to the user. One way to lessen the confusion problem in Symphony is to create menus that switch between windows that display each section of a template. Another problem arises if you have many @functions and formulae that tie big sections of the spreadsheet together. If your template has this attribute, errors may become a serious problem unless you thoroughly test the template before you use it. That is, changing data in one section may have a detrimental effect on other sections if you are not careful.

Create "Sectional" Reports

Sometimes a little planning will permit you to design a template that serves multiple purposes. For example, you may be able to get day-to-day performance data for quality control purposes from the same template that holds experimental data. Or, if you work in a centralized assay laboratory that performs sample analyses from several laboratories, your need to issue individualized reports (that contain only the data that a submitting lab is interested in) can be met by laying out a template that can be individualized. (You can use the /*W*orksheet ***C***olumn ***H***ide command (Lotus 1-2-3) or /*W*idth ***H***ide command (Symphony) to hide data from other submitting laboratories and just print the appropriate data for the lab in question. Lotus also allows ranges to be hidden. Hiding ranges is particularly useful when data are surrounded by data that need to be displayed. For Lotus 1-2-3, the command sequence is /***R***ange ***F***ormat ***H***idden. For Symphony, the command sequence is /***F***ormat ***O***ther ***H***idden.)

Use Easily Understood Labels

The labels that you use in a spreadsheet should be as easy for everyone to understand as possible. Use as little jargon as you can get away with. Bear in mind that some of the analysts that will use your spreadsheet may have minimal training and what may make sense to you, may not make sense to them. If you use non-descriptive terms, it may confuse them and lead to errors. This misunderstanding, at the very least, wastes time. It may also cause serious errors in data input and thereby lead to misinterpretation of an experiment.

Create and Erase Cells for Maximal Speed

The slowest process during data collection is the process of creating new cells in a template. Once created, the process of filling the cells with data is relatively quick. Likewise, overwriting old data in cells with new data is a relatively slow process. The fastest way to place data in cells is to have the cells already created, but empty. You can create this setup by forcing Lotus to create the cells at the time that you make your original template.

Thus, you would fill the cells with "dummy" data and then erase the data. Erasing existing cells clears the previous data, but leaves the cell's memory space and attributes intact. This method is faster than creating and filling new cells and is also faster than replacing existing data in cells. Therefore, if you need a fast data acquisition rate for an instrument, use the /***D***ata ***F***ill (Lotus 1-2-3) or /***R***ange ***F***ill (Symphony) command to fill your template with numbers. Then, erase the numbers and use the empty cells to receive data.

However, do not charge right out and create all of the cells in a spreadsheet. Even though the cells do not contain data, they still use up memory, meaning that you are likely to run out of memory and you may actually be slowing the performance of your spreadsheet if you create too many cells. It also means that you will be storing bigger files on your disk. Therefore, create only cells that you plan to use in your template and scratch-pad.

PROGRAM DESIGN TIPS

Each macro program that supplies functionality to LIM systems typically contains the following sections:

- User input to obtain specifications for record(s) to be imported or exported.
- User input to enter or edit data within the record.
- Formatting of data to conform to the specifications used in database definitions.
- Selection of the primary index file(s) to use.
- Determination of existing record numbers or creation of new records.
- If data are exported, retrieval of data from a template and transfer to a database.
- If data are imported, retrieval of record number-specific data from a database and transfer of data to a template.

Thus, there will be common functionality among your programs. This common functionality is best supported by creating generic subroutine utilities and using the subroutines as building blocks within macro programs tailored to specific tasks to be completed.

The following subsections highlight important topics to consider when creating macro programs and generic subroutine utilities.

Passing Arguments to Subroutines

For each argument that you pass to a subroutine, you must also specify a cell to store the value being passed and how the arguments being passed are to be interpreted. These designations are accomplished with a DEFINE statement at the beginning of the subroutine.

The ":VALUE" portion of a DEFINE statement represents a significant difference in the way in which subroutine calls and Lotus macro commands and @functions operate.

With Lotus macro commands and @functions, you can supply cell names, formulae, or actual values as arguments. With subroutine calls, you must supply the type of argument specified in the DEFINE statement. In general, the ":VALUE" suffix instructs Lotus to evaluate the argument before storing the value. That is, Lotus considers the argument to be a number, cell address, or formula. If you omit the ":VALUE" suffix, Lotus stores the argument as a label, exactly as it appears in the subroutine call. Conversely, if you use the ":VALUE" suffix and try to pass a string to the subroutine, an error would occur.

To circumvent this problem when you want to pass strings, you can place the strings in token @functions to fulfill the number, cell address, or formula requirement. @UPPER, @TRIM, and @CLEAN @functions make no substantive changes on strings containing no spaces. @UPPER is independent of spaces.

The choice of whether to use the ":VALUE" suffix depends on the intended use of the subroutine and argument. For example, if a subroutine will be called repeatedly during execution of {FOR} loops and its argument will change each execution, use the ":VALUE" suffix. If, on the other hand, the value of its argument will not change, do not use the ":VALUE" suffix. Arguments that usually do not change are index and database file names.

The number of arguments passed to a subroutine must match the number of arguments in the DEFINE statement. Otherwise, Lotus issues an error message. If you omit the DEFINE statement, no values are transferred, meaning that the subroutine cannot use any arguments specified in the subroutine call.

To summarize the importance of the above discussion, care must be maintained when using subroutines. The following are two guidelines to follow when passing arguments to subroutines:

- Be careful when passing arguments to subroutines. Arguments ***MUST*** be in the correct format. Most problems are due to using a string argument when a value is needed, and vice versa.
- Be careful to check the spelling of each argument before using subroutines.

@BASE @Function Arguments

Arguments used in @BASE @functions can be field names, record numbers, or values. The following are some guidelines for arguments:

- Do not include spaces in @BASE @functions.
- Arguments which refer to strings must be enclosed in single or double quotes.
- Do not use quotes for cell references, range references, and numbers.
- Arguments must be separated by commas.

Post Data When Using Macros to Control the @BASE Menu

Using "FP" commands when controlling the @BASE menu from macro programs is critical. FP is an @BASE menu command that performs a ***F***ile ***P***ost. Without this command, you would lose any changes that you have made to database files as soon as you leave the @BASE menu system. This peculiarity is due to a difference in the way that @BASE is controlled in manual versus macro mode. In manual mode, changes are automatically posted when you leave the menu system. When the @BASE menu is controlled by a macro, changes are not posted unless you specifically implement a post.

Also, because @BASE uses buffers to do reads and writes to disk, there may be data in buffers that have not been written to disk. If power was turned off, it would result in missing data. ***F***ile ***P***ost is similar to closing and reopening a database file, except that it retains any current criteria.

Select the Primary Index

If there are several index files open for a database file, you must specify which index file is primary. After a primary index file is selected, searches of the database will use the index key expression and key values in the primary index file. If you use the wrong index file as primary, searches will be ineffectual. For example, in Chapter 5 there were two index files created for the RESULTS.DBF database (RES_BC.NDX and RES_TRIO.NDX). If you tried to search the index file by barcodes and RES_TRIO was primary, no matches would be found.

Therefore, you should select the primary index files near the beginning of all macro programs. Although this selection may not be necessary in all instances, the action should be taken anyway. By specifically stating the primary index files, programs will be easier to interpret and independent of other, previously executed, programs.

Observe Differences in Numbers and Strings

Most operations that transfer data to databases have strict rules on whether values used are strings or numbers. Therefore, be very careful when using numbers. Sometimes numbers need to be converted to strings and/or padded before use. Other times they must be numbers. Input for lot numbers, barcode numbers, etc., tend to be user-dependent. Some users may use numbers instead of strings. {MK_STRNG} is necessary to ensure strings where appropriate.

Pad Strings Used as Key Values

You must pad strings when using index key expressions that have multiple or date fields. Padding adds spaces into strings so that the strings conform to the index key expression of the index file. Padding should ***NOT*** be used for fields specified as numeric when they not part of index key expressions containing multiple field names.

When an index key expression involving multiple fields or date fields is used, conversions of numbers to strings and concatenation of the strings are performed by @BASE before the key value is placed into the index file. A concatenation process "adds" character strings together. When a search of the index is performed, @BASE compares a search string for the record to be found to the key values in the index file. When an exact match is found, the record number is returned. However, there is a very important part of the index key expression concatenation process that you need to be aware of. When concatenation is performed on strings shorter than the field length specified in the database definition, @BASE adds spaces to the string to extend the length of the string to conform to the field definition. Because an exact character-for-character match is required, reliable searches would not occur unless the appropriate number of spaces were reliably added. This process is called ***padding***. The {PAD} subroutine in Chapter 8 automates the padding process.

An important consequence of padding is: If numbers are used in index key expressions involving multiple fields, use the {PAD} subroutine to convert to strings and pad the appropriate number of spaces. ***DO NOT*** use @STRING to convert numbers to strings and then assemble keys containing the strings.

Start by Finding Record Numbers

Before a record can be updated or the contents of a record's fields can be retrieved, the record's number must be made available. Similarly, the first occurrence of a record must be identified before @NDXNEXT and @NDXPREV @functions can be used. {GET_REC} supports these requirements and places the record number into the RECRD cell.

@Function vs. Menu-based Data Transfer

The {KEY_IMP} subroutine is used for importing tables of data from database files. {KEY_IMP} is much faster than using {GET_FLD} subroutines to obtain individual data fields from databases and then transferring the data to a template using {PUT} commands. This speed difference is due to the relatively slow performance of {PUT} commands and is especially noticeable when importing large amounts of data. However, {KEY_IMP} is more restrictive than {GET_FLD} because the data are placed in a table.

Importing data overwrites everything in the cells that fall within the output range. Be sure to leave enough space below the row of import definition labels to account for the number of records being imported.

Use Tables to Define Field Names and Offset Positions

Use tables to provide template column and row offsets for {PUT} commands and @INDEX @functions. Your tables should also contain field names for efficient transfer of data between databases and templates. Tables will make your programs more organized, more "readable," easier to follow, and easier to change if the need arises.

Test for Existing Records

It is very important to use {IF} commands prior to calling generic subroutine utilities based on @BASE menu commands. These {IF} commands are necessary to eliminate program termination when no records are found for the key value.

Concatenating Using {LET} Commands

When using {LET} commands to concatenate, precede the equation with a plus sign (" + ") if strings within quote signs are being used.

Specify Macro Names for {DISPATCH} Commands

Position {LET} commands before each {DISPATCH} command in your program to specify the macro program to branch to when the {DISPATCH} command executed.

Configure Your Programs to Be Independent of Lotus Release

It is important to use auto-executing macros to

- determine the Release of Lotus software being used,
- modify cells (based on Lotus Release) to achieve program portability, and
- activate menu systems.

By doing so, you will ensure portability of your program so that it can be shared with colleagues.

Use Named Cells and Ranges

Use named ranges and cells whenever possible. This rule is especially important for {PUT} and {LET} macro commands. Using named ranges will make your program easier to read and follow. Having a name rather than a range or cell address becomes very important in large spreadsheets. Perhaps more importantly, if you change your template and have cell addresses in macro commands, you will have to go through your program and change the addresses. However, if you change the address specifications of a named range and have used just the range name in a {PUT} or {LET} command, you would not have to edit the command.

Use Meaningful Names for Ranges and Macros

The names that you use should be as meaningful as possible. Don't get too fancy with names. Try to use names that relate to the processes that you are trying to carry out. Also, try to use names that remind you of the macro's or named range's purpose. Don't use names that look like cell addresses (e.g., A1); don't use any of the Lotus special keynames, keywords, or macro command names (e.g., PUT, QUIT); don't use spaces and/or symbols within the name; and do not exceed 15 characters. Sometimes you can break these rules, but most of the time you cannot.

Comment!

Add comments to your programs for one VERY important reason: Comments make it easier for someone (including yourself) to read and understand your programs. Just try to return to a program after several weeks and attempt to remember what you were trying to accomplish in a certain section of the program! Or try to remember a pitfall that you thought about while you were programming. If you document a program section at the time that you write it, the program would be at its freshest in your mind and the comments will make the best sense. In other words, comments will be more "readable" and it will be easier to correct or modify the program, if necessary.

Commenting your program also helps to clarify your own concept of what a program does and often exposes design flaws. ***When commenting, try to be brief. Remember, memory utilization is at a premium in LIM system applications. @BASE add-in application programs, macro programs, indexes, and spreadsheet data tend to consume large amounts of memory. This consumption means that you must use the remaining memory efficiently. Character-for-character, comments use the same amount of memory as program lines. Therefore, be concise in your comments to try to conserve memory.*** Because the programs in this book are oriented toward teaching you programming concepts, the programs tend to be over-commented. In most applications, the commenting would be less, and only comments of strategic importance would be used.

Provide a Program Synopsis

Another important piece of documentation is to place a brief synopsis of the function of the program at the beginning of the program. Because macro execution begins at the cell that corresponds to the macro's name, any text above the cell will be ignored. Your synopsis should contain the objectives of the program, instrument(s) controlled, type of testing performed, samples tested, databases influenced, etc.

Add a Development Tracking Table

Keeping track of the version of software that is being used is very helpful when a problem arises. It is invaluable for debugging purposes. If you have a table that tracks the history of program/template changes that were made to the spreadsheet, you can easily keep track of pertinent information. This table should be at the end of the macro program and should include the name of the person who developed the program, the date it was developed, the name of the individual(s) who modified it for each revision, and the revision date(s).

Prepare a Named Range Table

Prepare a Named Range table in your spreadsheet to ensure that you have created all of the ranges that you need for your program. More importantly, carefully review the cell specifications for each range name to make certain they are correct. Incorrect cell specifications are a common source of program error.

Structure the Program for Readability and Testing Ease

Testing a macro program can require as much time as programming it. If you invest the time to create a readable and well-organized program from the start, testing will be minimized.

Program in Sections

Macro programs should be constructed in sections, with each section being a macro subroutine. Don't try to make one large program. Large programs tend to be hard to follow, limited in their flexibility, hard to test, and obscure when pinpointing the causes of problems. Moreover, if your program is modular you can usually call the modules from several places in your program, thereby decreasing the number of programming lines.

Test as You Go

It is better to test macro subroutines as you create them rather than waiting until the entire program has been completed. If you wait until the end, there will be many subroutines and program lines, which will make it more difficult to isolate the errors. However, if you test as you build, you can verify that each subroutine works correctly before you add another layer of complexity and/or perform the final test that checks the entire program.

Test in Sections

The following advice is important and strongly recommended! If your program has several macro subroutines and/or your template has several sections, it is preferable to test each subroutine and section independently. Then, when everything is confirmed to work according to design, check to see if the parts interrelate correctly. It is very easy to overlook discrepancies in a subroutine or section if you attempt to check too much all at once. It is also possible that compensating errors will escape detection when the entire spreadsheet and macro program are checked at once.

Thoroughly Test the Final Program and Template

A template and macro program that have not been thoroughly tested ***under actual experimental conditions*** should not be used to report results. Important decisions may be based on a report that looks formal, but contains erroneous results. You can compare the results of the macro program with your manual method, the instrument's display, etc. You should compare the results of the calculations in your template with the results that you obtain from a calculator, etc.

Date-Stamp the Experiment

Date-stamp the experiment from the macro that collects data. Use a {LET} command and the @NOW function. This convention will ensure there is an accurate record of when the experiment was actually performed. If several experiments are being performed on the same day, you may also want to time-stamp each experiment.

Use Menus

Use custom menus whenever you can. These menus will allow your users to interact with a program as easily as possible. This way, your program will able to be used by personnel with a limited amount of specialized knowledge.

Minimize Screen "Flicker"

After you have worked with a macro program for even a short period of time, you will become tired of the "flickering" that takes place as the macro executes. "Flickering" is caused by the panel and spreadsheet updating during program execution. This "flickering" can be quite annoying to a user and will also slow execution of the program. Therefore, use a {WINDOWSOFF}{PANELOFF} combination to freeze the display.

Provide the User with Status Reports

It is important to inform your users of the status of your program as it executes. This notification is especially important if your program takes a relatively long period of time to complete its tasks. Under these circumstances it may appear to the user that something has gone wrong. The most convenient way to provide a status report is with the {INDICATE string}{PANELON}{PANELOFF} set of commands.

Don't Start with Nested @Functions and/or Macro Commands

It is better to start with individual small formulae than to jump right into long formulae with many levels of nesting. This design approach allows you to check out each component separately. Once you have individual formulae working correctly, you can combine them into the final (complex) formula.

Use {CALC} to Prevent User Confusion

The {CALC} command forces a recalculation of all formulae in a spreadsheet and updates (re-displays) the most current data. When a macro program executes, it does not necessarily recalculate the formulae in a template. Placing a {CALC} at the end of a macro is very important to avoid user confusion. More importantly, if a user forgets to recalculate the spreadsheet before printing it, the results may not be correct, because they may not reflect new values that have been added to the template. To prevent the possibility of catastrophe, you should get into the habit of placing a {CALC} command at the end of all of your macro programs and, more importantly, before any automated printing is performed by your program.

SPREADSHEET PROTECTION: DESIGN TIPS

Once you have created your templates and macro programs and have begun generating data, your spreadsheets will represent a considerable time investment. You will want to protect this investment in any way that you can. Below are some things that you can do to try to prevent devastating errors that can ruin your work. There are also some things that you can do to prevent unauthorized viewing of confidential information.

Restrict File Access

You can assign passwords to spreadsheet files so that a password entry is required before a spreadsheet can be retrieved. That is, anyone who wants to retrieve the spreadsheet will need to enter the correct password before Lotus will allow access to the spreadsheet. This protection permits you to restrict file use to only those people in the organization who are authorized to work with the file. This password will restrict access to the file, but once the file is retrieved the user will be able to make changes to the entire file. To add a password to a file in Lotus 1-2-3, store the file with the /***F***ile ***S***ave filename ***P*** command (where filename is the name of the file that you wish to save). After the filename, type one or more spaces followed by a lower-case or upper-case P. In Lotus Symphony, the [SERVICES] ***F***ile ***S***ave filename ***P*** command is the appropriate sequence. Lotus will ask you for a password before it saves the file.

Cell Protection

You can ensure that less experienced users cannot accidentally ruin a spreadsheet by placing an entry in, erasing, or deleting the wrong cell(s). You can also ensure that a program will not be overwritten. Lotus 1-2-3 and Symphony both have protection features that allow you to choose which cells that you want to protect and which ones that you want to allow updates to. Once a cell is protected, it cannot be changed by a user or macro. Lotus establishes a protection default status of OFF for all cells in a new spreadsheet. This status allows changes to any cell. To use cell protection, you must first enable protection for the entire spreadsheet and then turn it off for the cells for which you want to allow changes. In Lotus 1-2-3, issue the /*W*orksheet *G*lobal *P*rotection *E*nable command. In Symphony, issue the [SERVICES] *G*lobal-protection *Y*es command. This sequence turns protection on for all cells. Next, issue the /*R*ange *U*nprotect command in Lotus 1-2-3 or the /*R*ange *P*rotect *A*llow-changes command in Symphony and specify the range of cells for which you will allow changes. At a minimum, unprotect the data input sections of your template ***and the scratch-pad section of your macro program.*** As a general rule, you can leave cells that have formulae in them protected and they will update as usual; but you MUST unprotect all cells that macro commands such as {READLN}, {PUT} and {LET} interact with.

Hide Confidential Information and the Macro Program

If your template contains confidential information or if you just don't want users to see your program, you can hide columns in a spreadsheet. In Lotus 1-2-3, use the /*W*orksheet *C*olumn *H*ide command. In Symphony, the command is /*W*idth *H*ide. Symphony also has another way to hide areas of a spreadsheet. It is the /*F*ormat *O*ther *H*idden command and it will hide the ranges that you specify. In this way, you are not limited to just columns.

Backup

Store your spreadsheet often while developing it. That way, if you have a power problem or computer lockup, you won't lose the work that you have already completed. Plan for future disasters, too. If you have a large spreadsheet and/or complicated macro program, you will also want to store several copies ***without data*** on separate floppy disks and store the disks in separate locations. You should print out the template and macro program and store them in safe places as well.

PROGRAM TROUBLESHOOTING TOOLS

Whenever you are confronted with a problem in a program, the first thing that you need to do is to get as much information surrounding the problem as you can. That is, you should try to characterize when the problem happens, where in the program it happens, and the consequences of the problem. The more information that you have, the better your chances of being able to correct the problem.

There are a number of programming tools that you can use to increase the amount of information available to you. This section describes some of the more useful ones.

Update the Named Range Table

One of the most frequent causes of program problems is ranges with incorrect cell specifications. There are a number of situations that cause range specifications to change without your knowledge. Therefore, anytime that you experience a problem, it is a good policy to update the Range Name table and review it to make sure it is correct. Review the named ranges to ensure that the ranges are accurate and not overlapping (two or more range names with the same starting and ending cell designations). A sudden slowing of the program or range definitions that appear to change for no reason are indicative of overlapping ranges.

To make a Range Name table, issue the /***R***ange ***N***ame ***T***able command. When prompted for the location of the table, press [ESC]ape and move the cell-pointer to a cell outside the area being used for the template. Press [RETURN] and the table will be created.

If problems are substantial, you may want to consider deleting all of the range names and re-creating the named ranges. Sometimes moving ranges causes overlaps in more than one set of ranges.

When you are finished with the table, delete and re-insert the columns being used by the table. This final task will free up a substantial amount of memory for your program.

Prepare a Template for Spreadsheet and Database Data

One very useful tool is to create a template at the bottom of a program so that you can compare values in your database with the ones in your templates and scratch-pad. The most common database programming problems involve the inability to obtain accurate record numbers and subsequently transfer data between database fields and templates. A temporary template located near the program will allow you to reveal the source of the problem.

An example template is shown in Figure 10-1. A template of this nature contains cell references for the spreadsheet and @DB and @NDX @functions for the database file. The template shown is an actual tool that was instrumental in locating and correcting a problem during the development of the EXP_BCH program shown in Figure 9-18. The following table describes the formulae contained in the cells of Figure 10-1:

Cell	Formula
DP63	+INTERCEPT
DP64	+SLOPE
DP65	@INDEX(STAT_TB,@INDEX(S_TAB,1,6),@INDEX(S_TAB,2,6))
DS63	+RECRD
DS64	+TRAP
DS65	+TECH_COMS
DP68	@DBFLD("BATCH","CURV_A",RECRD)
DP69	@DBFLD("BATCH","CURV_B",RECRD)
DR68	@DBFLD("BATCH","STD_C",RECRD)
DR69	@DBFLD("BATCH","TECH_COMS",RECRD)

```
--------DN--------DO-------DP--------DQ--------DR-------DS-------DT--------
61
62                  /*SPREADSHEET TEMPLATE TEMPORARY VIEW PAD*/
63        INTERCEPT94.51842          RECRD          51
64        SLOPE    1.494878          TRAP            1
65        STD_C    0.016751          COMS     24-HOUR INCUBATION
66
67                 /*DATABASE TEMPORARY VIEW PAD*/
68        CURV_A   94.51842 STD_C    0.016751
69        CURV_B   1.494878 TECH_COMS 24 HR INCUBATION
70
```

Figure 10-1

Evaluate Equations Within {PUT} and {LET} Commands

Another very useful tool is to copy {PUT} and {LET} commands containing questionable formulae to open cells of a spreadsheet and use the [F2] (edit) key to enter the edit mode. Next, delete all portions of the command not pertaining to the formula. Then, press [RETURN] and evaluate the value returned by the formula for accuracy.

@DB and @NDX @functions have one quirk that you need to be aware of. Under some conditions when these @functions are in cells, they do not update when the spreadsheet updates. For this reason, you must use the [F2] key and press [RETURN] before evaluating the returned value.

If possible, templates created for the purposes of the previous subsection should be created using formulae of this origin.

Turn the Windows and Panel Back on

Another action that you should take when you are having a problem is to place a {WINDOWSON}{PANELON} combination at the beginning of your program and temporarily remove all of the {WINDOWSOFF}{PANELOFF} commands. This change will allow you to see exactly what is happening as your program executes. You should also place {CALC} commands throughout the portion of the program that is giving the problem, thus ensuring that @functions recalculate and that the data you are viewing accurately reflect what is actually happening in the program.

Slow the Program Down with {WAIT} Commands

Most macro programs execute very quickly. You can slow the execution of a program as much as you want by placing {WAIT} commands throughout the program. In this way, you can view the panel and template or scratch-pad to try to get a clue as to what is actually happening. Again, use {CALC} commands to make sure that you are getting frequent updates of formulae. Alternatively, use the Lotus STEP mode to view updates. (See below.)

Single Step Through Your Program

Lotus 1-2-3 and Symphony both provide a STEP mode for macro testing. In Lotus 1-2-3, you can activate the STEP mode by pressing [ALT] F2. In Symphony, press [ALT] F7. Start the macro in the usual way (by pressing [ALT] and macro's character name; or if you have Symphony, use F7 and the name of the macro). Thereafter, each time you press the [SPACE BAR], another macro command will be executed. This command allows you to trace each step of a macro's execution and helps you to locate the instruction that is causing problems. When you are finished, press [ALT] F2 (Lotus 1-2-3) or [ALT] F7 (Symphony) again to return to normal operation.

Symphony 2.2 takes the STEP mode one important step further: Release 2.2 displays commands as they execute. This display is very useful because you can actually trace the flow of a program, ensure that {IF}, {DISPATCH}, {BRANCH}, etc., commands are executing correctly, and compare the flow to the flow that you would expect to see.

Track the Program

You can track the execution of your program at close to full speed by placing {LET} commands throughout the program. That is, alternate your program lines with {LET} commands, each {LET} command placing a unique number into a status cell. By viewing this status cell, you can determine where you are in a program. This tool is very useful if your program is locking up. You can use the information received from this technique to pinpoint exactly where the program has stopped executing. That is, you can backtrack from the number in the cell to the last {LET} command that executed when you had to press [CONTROL][BREAK] to abort the locked-up macro.

Use {INDICATE}

Alternatively, you can use the status indicator box. By placing sequential numbers in {INDICATE} commands placed throughout your program, you can trace the execution of your program. If you use this technique, make sure that a {PANELOFF} command has not been used prior to the {INDICATE}. The {INDICATE} command requires the panel to be "on" for it to update.

WHAT TO LOOK FOR

This section provides some suggestions on what to look for if you have a problem. Many of the items on the list use information that you can obtain using the troubleshooting tools provided previously.

- Before you check anything else, check the spelling of all statements in your program. Most errors are due to incorrectly spelled arguments passed to subroutines, macro commands, and @functions.
- Check the arguments being passed to your subroutines. With subroutine calls, you must supply the type of argument specified in the DEFINE statement. Ensure that strings and values passed correspond to the type in the DEFINE statement of the subroutine. Arguments ***MUST*** be in their correct format. Most subroutine problems are due to using a string argument when a value is needed, and vice versa.
- If a numeric field in a database does not receive data, check the field width. @BASE will not place a value into the field unless there is enough room for all of the integers. You should also check the data type to ensure that it has not been inadvertently defined as character.
- Incorrectly defined field widths are also common problems when creating joined files. If the widths (and types) of the fields used to join the primary and lookup files are not identical, records in the lookup file will not be found.
- If a cell receives a string equivalent to a formula in a {LET} or {PUT} command, the situation is a clear sign that

    - a range name within the formula does not exist.
    - the formula or range name is incorrectly spelled.
    - a ":VALUE" suffix was omitted from the DEFINE statement in a subroutine.

- Check the number of arguments being passed to the subroutines being called by the program. The number of arguments passed to a subroutine must match the number of arguments in the DEFINE statement. Otherwise, Lotus issues an error message. If you omit the DEFINE statement, no values will be transferred, meaning that the subroutine cannot use any arguments you specify in the subroutine call.
- If your @BASE @functions are returning ERRs, check the arguments in the @functions. They must conform to the following standards:

    - Do not include spaces in @BASE @functions.
    - Arguments which refer to strings must be enclosed in single or double quotes.
    - Do not use quotes for cell references, range references, and numbers.
    - Arguments must be separated by commas.

- If you are having a problem with program flow, check to ensure that there are {LET} commands in your macro programs that specifically change any pre-existing labels in the SUB_PTR cell to the names of the macros you want your programs to branch to when their {DISPATCH} commands execute.

- Most operations that transfer data to databases have strict rules on whether values used are strings or numbers. Make certain that the correct data type is being used. You must be especially careful when using numbers. Sometimes numbers need to be converted to strings and/or padded before use. Other times they must be numbers.
- If you use an incorrect data type during searches, you will not find records. If numbers are used in index key expressions involving multiple fields, ensure that the numbers are converted to strings and padded to the appropriate number of spaces. ***DO NOT*** use @STRING to convert numbers to strings and then assemble keys containing the strings. However, if a number is used for a ***single*** field index key expression, the number ***cannot*** be a string.
- If your program cannot locate records in a database file, another item to check is the primary index file. If there are several index files open for a database file, you must specify which index file is primary. If you use the wrong index file as primary, searches will be ineffectual because the wrong index key expression for the key value will be used.
- If you receive an "Unknown Field Type" message, the message sometimes means that you have the wrong spelling for one of the field names in the spreadsheet. However, most often the message is received because field names in the specification range have all upper-case characters for one or more virtual fields. For virtual fields, you must use lower-case characters.
- If, when exporting a table of records, either not all of the records or empty records appear in the database, check the size of the export range. The range must start at the row containing the field names and extend down the data to be exported. The range cannot be open-ended. If the range is larger than the table of data to be exported, empty records will be created. If the range is smaller than the table of data, not all of the records will be exported.
- If you are having trouble manipulating records, check whether the {GET_REC} subroutine is being called to locate the record number prior to data manipulation. {GET_REC} ***MUST*** be used prior to calling {UPDATE}, {GET_FLD}, {GET_NEXT}, {GET_PREV}, etc.
- If you export the same record(s) more than once, expecting to overwrite archived record(s) but find the record(s) being split (i.e., parts of the data appear in different record numbers), then one of the following is the probable cause:

    - A macro program was used and a ***F***ile ***P***ost was not used after exporting the data.
    - The key value was not padded.
    - A numeric key value was padded.

- A sudden slowing of a program or range definitions that appear to change for no reason are indicative of overlapping ranges (two or more range names with the same starting and ending cell designations).

- If you receive an "Index file out of synch - reindex" error message, there are record numbers in the database file that are not in the index file. This problem can occur if changes are made to the database file while the index file is closed or if ***F***ile ***P***ost was not used after the @BASE menu system was accessed using a macro program. To correct the problem, the index file must be re-indexed to ensure proper operation. To do so, select ***O***ptions ***I***ndex (***F***ile) ***R***eindex from the menu. Select an index file to rebuild and press [RETURN]. Select ***E***xecute to rebuild the index file or ***C***ancel to stop the process.
- If you have a problem with concatenating strings with {LET} commands, check to make sure that a plus sign is being used when strings within quote signs are being concatenated.
- If, when using your system, you receive a "Cannot open file," "Locked file," or "Cannot create file" error message, check your CONFIG.SYS file for a FILES command to ensure that the command is present and specifies at least 20 files. If FILES=20, then a maximum of about 14 files are available because Lotus and @BASE normally maintain up to about 6 open files. Because your system may utilize up to 14 open files, use your word processor to change the FILES statement to at least 20 and re-boot your computer. (The maximum number of files is 255. However, DOS allows no more than 20 files to each software application. So, increasing the number to more than 20 files will not increase the number of files available to Lotus unless you have Terminate-Stay Resident (TSR) programs running that also have open files. If your system exceeds the 14 open file limit, then you will need to review the system and try to combine database files or close open index or database files to decrease the number of open files to an acceptable limit.)

SUMMARY OF WHAT HAS BEEN ACCOMPLISHED

This chapter contained summary information aimed toward efficient and reliable database, template and program design. Perhaps more importantly, the chapter provided tools that you can use to identify and correct the cause of problems should they occur.

WHAT'S NEXT?

The remainder of this book contains a number of useful appendices. The first three appendices provide summaries for Lotus arithmetic, relational, and logical operators, @functions, and macro commands. The last two appendices contain summaries for @BASE @functions and operating functions.

Arithmetic, Relational, And Logical Operators

Lotus 1-2-3 and Symphony use a set of "operators" to depict arithmetic operations. For example, the + operator causes the two values flanking it to be added together.

Lotus has a well-defined hierarchy that is used to determine the progression in which operators will be used. For example, consider the equation:

value = 10.0 + 35.0 * slope / factor

This equation has an addition, a multiplication, and a division. Which operation takes place first? Is 10.0 added to 35.0, the result of 45.0 then multiplied by slope, and that result then divided by factor? Or is 35.0 multiplied by slope, the result added to 10.0, and that answer divided by factor?

Clearly, the order of execution for various operations can have a profound influence on the final value. To guard against chaos, Lotus needs unambiguous rules for choosing what to do first. Lotus sets up a pecking order for its operators. This pecking order is called "precedence" and each operator is assigned a precedence level.

The table shown in Figure A-1 gives the precedence for each Lotus operator. In the table, division and multiplication have higher precedences than addition and subtraction, so division and multiplication are performed first. If the precedence of two or more operators in a statement is the same, they will be executed according to the order in which they occur in the statement (left to right). If you are unsure of the precedence in an equation (or if you want to change it), use parentheses. Whatever is enclosed in parentheses is performed first. Within parentheses, the rules defined in the table apply.

In addition to the arithmetic operators, Lotus has a category of operators that can be used for making comparisons. These operators are called "relational" operators. Each

Operator	Definition	Precedence
^	Exponentiation	7 (highest)
-	Negative	6
*	Multiplication	5
/	Division	5
+	Addition	4
-	Subtraction	4
=	Equal	3
&	String Concatenation	1 (lowest)

Figure A-1

of the relational operators compares the value at its left to the value at its right. The relational expression evaluated from an operator and its two operands has the value 1 if the expression is true and the value 0 if the expression is false.

Relational operators are most commonly used for testing the relationship of two expressions in an {IF} macro command or an @IF function. The table in Figure A-2 gives the definition and precedence for each relational operator. The precedences are on the same scale as arithmetic operators. Because the precedences are lower than those for arithmetic operators, arithmetic operations occur before relational operations are evaluated.

Operator	Definition	Precedence
<	Less than	3
<=	Less than or equal	3
>	Greater than	3
>=	Greater than or equal	3
<>	Not equal	3

Figure A-2

The final category of operators are "logical" operators. Logical operators normally take relational expressions as operands. In this context, logical operators are useful for combining two or more relational expressions. The table in Figure A-3 gives the definition and precedence of each logical operator. Because the precedences for logical operators are lower than the precedences for both arithmetic and relational operators, logical operators are evaluated last.

The logical expression evaluated from an operator and its operands has the value 1 if the expression is true and the value 0 if the expression is false. The following are brief explanations of each logical operator:

Operator	Definition	Precedence
#NOT#	Logical NOT	2
#AND#	Logical AND	1
#OR#	Logical OR	1

Figure A-3

#NOT#: The expression is true if the operand is false, and vice versa.

#AND#: The combined expression is true if *both* operands are true, and false otherwise.

#OR#: The combined expression is true if one *or* both operands are true, and false otherwise.

Logical expressions are evaluated from left to right. The evaluation stops as soon as the expression is rendered false. The following examples illustrate the use of logical operands:

6>2#AND#3>1	is true
#NOT#(6>1#AND#10>4)	is false
6>7#OR#6>1	is true

@Function Summary

Special-purpose Lotus formulas are called "@functions". There is a very large number of Lotus @functions. Each @function will extend your calculating power far beyond simple arithmetic and text-handling operations. There are several broad categories of Lotus @functions: mathematical, statistical, string, special, logical, and date/time.

The following is a list of the @functions used in this book and some others that you will find useful. The definitions are tailored to the primary purpose of the book--the creation of Lotus spreadsheets which have powerful data reduction templates and macro programs.

Mathematical @Functions

`@ABS(number)`
Absolute value of *number*.

`@ACOS(number)`
Arc cosine of *number*.

`@ASIN(number)`
Arc sine of *number*.

`@ATAN(number)`
Two-quadrant arc tangent of *number*.

`@ATAN2(number1,number2)`
Four-quadrant arc tangent of *number2/number1*.

`@COS(number)`
Cosine of *number*.

`@EXP(number)`
The number e raised to the *number* power.

`@INT(number)`
Integer part of *number*.

`@LN(number)`
Log of *number*, base e.

`@LOG(number)`
Log of *number*, base 10.

`@PI`
The number pi (approximately 3.14159).

`@SIN(number)`
Sine of *number*.

`@SQRT(number)`
Positive square root of *number*.

`@TAN(number)`
Tangent of *number*.

Statistical @Functions

`@AVG(list)`
Arithmetic average of values in *list*.

`@COUNT(list)`
Number of cells containing values in *list*.

`@MAX(list)`
Maximum of values in *list*.

`@MIN(list)`
Minimum of values in *list*.

`@STD(list)`
Standard deviation of values in *list*.

`@SUM(list)`
Sum of values in *list*.

`@VAR(list)`
Variance of values in *list*.

String @Functions

`@CHAR(number)`
Returns the ASCII/LICS character represented by *number*.

`@CLEAN(string)`
Removes the control characters from *string*.

`@CODE(string)`
Returns the ASCII/LICS code of first character in *string*.

`@EXACT(string1,string2)`
Tests whether *string1* and *string2* have exactly the same characters.

`@LEFT(string,number)`
Leftmost *number* of characters in *string*.

`@LENGTH(string)`
Total number of characters in *string*.

`@LOWER(string)`
Converts all upper-case letters in *string* to lower-case.

`@MID(string,startingposition,numberofchars)`
Extracts *numberofchars* characters from *string*, beginning at offset *startingposition*.

`@N(cell)`
Numeric value of *cell*.

`@RIGHT(string,number)`
Rightmost *number* of characters in *string*.

`@S(cell)`
String value of *cell*.

`@STRING(cell,places)`
Converts the number in *cell* into a string with *places* decimal places.

`@TRIM(string)`
Removes the leading and trailing spaces from *string* and compresses multiple spaces within *string* into single spaces.

`@UPPER(string)`
Converts all lower-case letters in *string* to upper-case.

`@VALUE(string)`
Converts a *string* that looks like a number into that number.

Special @Functions

`@CELLPOINTER(spec)`
Attribute of the cell currently highlighted by the cell pointer.

`@CHOOSE(selector,val1,val2,val3,...)`
Selects value based on its position in the list.

`@HLOOKUP(selector,rowrange,offsetnumber)`
Table lookup, comparing *selector* value with a row of values. The value of *selector* may be either a string or number.

`@INDEX(range,coloffset,rowoffset)`
Lookup based on position in *range*.

`@VLOOKUP(selector,colrange,offsetnumber)`
Table lookup, comparing *selector* value with a column of values. The value of *selector* may be either a string or number.

Logical @Functions

`@IF(criterion,value1,value2)`
If criterion is non-zero, *value1* is returned; if criterion is zero, *value2* is returned.

`@ISERR(value)`
If *value* has the value ERR, then 1 (TRUE) is returned; otherwise 0 (FALSE).

`@ISNUMBER(value)`
If *value* has a numeric value, then 1 (TRUE) is returned; otherwise 0 (FALSE).

`@ISSTRING(value)`
If *value* is a string, then 1 (TRUE) is returned; otherwise 0 (FALSE).

Date and Time @Functions

`@DATE(year,month,day)`
Serial number of specified date (Jan 1, 1900 = 1).

`@DATEVALUE(datestring)`
Serial number of date specified by *datestring*.

`@NOW`
Serial number of current moment.

`@TIME(hour,minute,second)`
Serial number of specified time of day.

`@TIMEVALUE(timestring)`
Serial number of time of day specified by *timestring*.

`@DAY(serialnumber)`
Day number of *serialnumber*.

`@HOUR(serialnumber)`
Hour number of *serialnumber*.

`@MINUTE(serialnumber)`
Minute number of *serialnumber*.

`@MONTH(serialnumber)`
Month number of *serialnumber*.

`@SECOND(serialnumber)`
Second number of *serialnumber*.

`@YEAR(serialnumber)`
Year number of *serialnumber*.

`@END(appendix)`
The end of this *appendix*. (Not a real @function.)

Macro Command Summary

A set of instructions in a format and language that Lotus can understand is called a Lotus "macro" program. The language that is used is called the "Lotus Command Language" and the instructions are called "macro commands".

There are six broad categories of Lotus macro commands. They are:

- **System commands**, which allow you to control the screen display and computer's speaker.
- **Interaction commands**, which allow you to create interactive macros that pause for the user to enter data from the keyboard.
- **Program flow commands**, which let you include branching and looping in your program.
- **Cell commands**, which transfer data between specified cells and/or change the values of cells.
- **File commands**, which work with data in DOS files or communications ports.
- **Menu commands**, which allow you to manage menus of your own design.

The following is a list, by category, of the macro commands used in this book. The definitions are tailored to the primary purpose of the book--the creation of Lotus spreadsheets which have powerful data reduction macro programs.

System Commands

{BEEP tone}
Sounds one of four different tones (1-4).

{INDICATE string}
Replaces the standard mode indicator message at the upper right corner of the display with the characters specified by *string*. Omitting *string* restores control of the mode indicator to Lotus.

{PANELOFF}
Suppresses updating of the control panel during macro execution.

{PANELON}
Restores standard updating of panel, canceling a {PANELOFF} command.

{WINDOWSOFF}
Suppresses updating of the display during macro execution.

{WINDOWSON}
Restores standard updating of display, canceling a {WINDOWSOFF} command.

Interaction Commands

{?}
Halts macro execution temporarily, allowing the user to type in data, move the cell-pointer, and access Lotus menus. Macro execution resumes when [RETURN] is pressed.

{GETLABEL prompt,cell}
Halts macro execution, displays a *prompt* for the user to type a line of characters, and stores the response as a label in the specified *cell*.

{GETNUMBER prompt,cell}
Halts macro execution, displays a *prompt* for the user to type a number or numeric expression, and stores the response as a number in the specified *cell*.

Program Flow Commands

{macro}
Calls *macro* as a subroutine.

{\R}
Calls *\R* as a subroutine. Same as pressing [ALT]-R.

`{BRANCH macro}`
Transfers control of the program to the location called *macro*.

`{IF condition}`
Conditionally executes the commands that follow the IF command in the same cell if *condition* evaluates to true.

`{ONERROR macro,messagecell}`
Branches to *macro* if an otherwise fatal error occurs. Optionally stores the error message that Lotus would have displayed in *messagecell*.

`{FOR counter,start,stop,stepsize,subroutine}`
Repeatedly executes the commands, beginning at the cell called *subroutine*, as many times as indicated by the values of *start*, *stop*, and *stepsize*.

`{QUIT}`
Immediately terminates macro processing, returning to manual keyboard control.

`{WAIT time}`
Causes macro execution to pause until the computer's clock time matches or exceeds *time*.

Cell Commands

`{BLANK range}`
Erases the cells within the specified *range*.

`{LET cell,number}`
Places *number* into the specified *cell*.

`{LET cell,string}`
Places *string* into the specified *cell*.

`{PUT range,column,row,number}`
Places *number* into the cell in *range* represented by offset coordinates (*column,row*).

`{PUT range,column,row,string}`
Places *string* into the cell in *range* represented by offset coordinates (*column,row*).

`{RECALC range}`
Recalculates the formulae in *range*, proceeding column by column.

File Commands

`{CLOSE}`
Closes a disk file or, more commonly, a communications port.

`{OPEN filename,mode}`
Opens *filename* (disk file or communications port) for read only (R), write only (W), append (A), or both read/write (M).

`{READ bytes,buffer}`
Reads the number of characters specified by *bytes* from a file or communications port and stores the characters as a string in the cell specified by *buffer*.

`{READLN buffer}`
Reads a line of characters (until a carriage return is encountered) from a file or communications port and stores the characters as a string in the cell specified by *buffer*.

`{WRITE string}`
Writes a string of characters (without a carriage return) to a file or communications port.

`{WRITELN string}`
Writes a line of characters (appending a carriage return) to a file or communications port.

Menu Commands

`{MENUBRANCH menu}`
Halts macro execution temporarily, branches to a customized menu whose instructions are found starting at the cell called *menu*, prompts the user to make a choice, then continues execution based on the choice.

`{MENUCALL menu}`
Similar to {MENUBRANCH}. However, processes *menu* and its associates steps as a subroutine.

Macro Commands for Special Keys

In addition to the above macro commands, there are commands which represent special Lotus keys. Cursor control commands allow you to specify the number of times your macro should press the named key. For example, {RIGHT 5} is equivalent to manually pressing the right arrow five times. The following is a table of these commands:

```
{EDIT}              {PGUP}
{GOTO location}     {PGDN}
{WINDOW}            {END}
{CALC}              {ESC}
{UP}                {BACKSPACE}
{DOWN}              {BIGLEFT}
{LEFT}              {BIGRIGHT}
{RIGHT}             {MENU}
{HOME}              ~(CARRIAGE RETURN)

      {SERVICES} or {S} (SYMPHONY ONLY)
```

@BASE @Function Summary

@BASE and the @BASE Option Pac contain @functions that establish connections between the spreadsheet and database files residing on disk. The most common use of the @NDX @functions supplied with the @BASE Option Pac is in the determination of record numbers. Once record numbers are determined, @DB @functions enable you to analyze, update, or perform calculations.

Each @BASE @function uses one or more parameters specifying database files, records, or fields. @BASE @functions can work on all records in a database, or just the records that are not marked for deletion and meet the active criteria. You can also set up criteria strings in certain @DB @functions that take precedence over or add to the active criteria for a given database.

Like @functions for Lotus, all of the @BASE @functions return a single value. For example, the @DBSUM @function returns a numeric value equal to the sum of the values in a specified field. The @DBFLD function returns the value in a specified field of a given record whether it is a numeric, string, or logical value.

@BASE @FUNCTION FORMAT

The general formats of the @DB and @NDX @functions are:

```
@DB[function name](argument1,argument2,...{criteria string})
@NDX[function name](argument1,argument2,...)
```

The *function name* tells @BASE which @function to perform. The ***arguments*** are the parameters used in the @function's operations or calculations. The first argument is usually the alias of the database file you want @BASE to perform the @function on. The remaining arguments can be field names, record numbers, or values. The ***criteria string*** is optional and can be used to select records from the database during the processing of @DB @functions. For example, the function,

```
@DBCNT("RESULTS","RUN_DT>33208#AND#RUN_DT<33238")
```

calculates the number of records with RUN_DT fields in the RESULTS database having dates falling within December 1990 and returns the number. The @function name is @DBCNT and the first argument is the alias, RESULTS. The second argument is the criteria string "RUN_DT>33208#AND#RUN_DT<33238".

Do not include spaces in @BASE @functions. The arguments, and any criteria, must be enclosed in parentheses. Separate arguments with commas and be sure to include double quotes around string arguments and criteria. If you use a criteria string with a string constant, enclose the string constant in single quotes and the entire criteria in double quotes. If the comparison value in the criteria string is either a numeric value, cell address, or named range, do not use quotes.

@BASE @FUNCTION ARGUMENTS

Arguments in @BASE @functions can consist of string values, numeric values, cell addresses, range names, nested @BASE @functions, or expressions.

CRITERIA IN @DB @FUNCTIONS

@DB @functions act only on records that meet the active criteria for the given database file. Some of the @DB @functions allow you to enter an ***optional criteria string*** as one of the arguments. If an optional criteria string is entered as one of the arguments, it takes precedence over the active criteria for the given database.

When multiple criteria are desired, ***logical operators***, similar in form and function to Lotus #AND# and #OR# logical operators, can be used. These operators allow you to temporarily "add" criteria strings.

The optional criteria are only logically added to the active criteria during the processing of @DB @functions and have no effect on the active criteria for the database file.

If you want to include active criteria (set through the @BASE menu system) to the criteria in the @DB @function, start the criteria argument of the @DB @function with a logical operator (#AND# or #OR#).

@BASE @FUNCTION PARAMETERS

The following terms represent the parameters used in @BASE @function arguments:

- *ALIAS* is the internal name @BASE uses to identify an open database file.
- *DATABASE* is the full name of a database file, including the extension.
- *FIELDNAME* is the name of a field in a specified database.
- *VALUE* is the value in a particular field.
- *RECNO* is a record number in a specified database file.
- *CRITERIA* is the optional selection criteria.

STATISTICAL @DB @FUNCTIONS

Statistical @DB @functions compute values based on numeric data in an open database file. Each @function returns a numeric value.

`@DBAVG(ALIAS,FIELDNAME,CRITERIA)`

Returns the average value of the numeric field FIELDNAME for all records in a database file ALIAS. Records must meet the active or optional criteria argument to be included in the computation. Records marked for deletion are not included if @BASE is set to ignore deleted records.

This @function can only be performed on a numeric field. However, dates are stored in the Lotus date format. Thus, you can retrieve an average date from a date field, but you must set up the cell receiving the value from the @DBAVG function with a Lotus long international date format.

`@DBCNT(ALIAS,CRITERIA)`

Returns the number of records in ALIAS. Records must meet the active and/or optional criteria to be counted. Records marked for deletion are not counted if @BASE is set to ignore deleted records.

`@DBMAX(ALIAS,FIELDNAME,CRITERIA)`

Returns the largest value of the numeric field FIELDNAME in the database file ALIAS. Records must meet active and/or optional criteria to be included in the computation. Records marked for deletion are not included if @BASE is set to ignore deleted records.

@DBMAX can be used on a date field to obtain a maximum date, but you must format the cell containing the @function in a Lotus long international date format to get a date as the result.

`@DBMIN(ALIAS,FIELDNAME,CRITERIA)`

Returns the minimum value of the numeric field FIELDNAME in the database file ALIAS. Records must meet active and/or optional criteria to be included in the computation. Records marked for deletion are not included if @BASE is set to ignore deleted records.

@DBMIN can be used on a date field to obtain a maximum date, but you must format the cell containing the @function in a Lotus long international date format to get a date as the result.

`@DBRECS(ALIAS)`

Returns the total number of records in the database file ALIAS, regardless of the active criteria and number of records marked for deletion.

`@DBSUM(ALIAS,FIELDNAME,CRITERIA)`

Returns the sum of the values in the numeric field FIELDNAME of the database file ALIAS. Records must meet active and/or optional criteria to be included in the computation. Records marked for deletion are not included if @BASE is set to ignore deleted records.

LOGICAL @DB @FUNCTIONS

Logical @DB @functions test data in an open database file for a specified condition and return a 1 (True) or 0 (False).

`@DBISACT(ALIAS,RECNO,CRITERIA)`

Returns a 1 if record RECNO in the database file ALIAS is not marked for deletion and meets the active and/or optional criteria. Otherwise, @DBISACT returns a 0.

If @BASE is set to ignore deleted records, @DBISACT returns a 0 if the record is marked for deletion. If @BASE is set to use deleted records, the record must simply meet the criteria for a 1 to be returned.

`@DBISDEL(ALIAS,RECNO)`

Returns a 1 if record RECNO in the database file ALIAS is marked for deletion. Otherwise, @DBISDEL returns a 0.

@DBISDEL returns a 1 for marked records regardless of whether @BASE is set to ignore deleted records.

`@DBISNA(ALIAS,FIELDNAME,RECNO)`

Returns a 1 if FIELDNAME of record RECNO in the database file ALIAS has no value. If the field has a value, the @function returns a 0.

A numeric field does not test true for @DBISNA if it contains a 0. Numeric fields must be blank to have no value.

@DBISSEL(ALIAS,RECNO,CRITERIA)

Returns a 1 if record RECNO in the database file ALIAS meets active and/or optional criteria. Otherwise @DBISSEL returns a 0. A record meeting the criteria will return a 1 even if it is marked for deletion.

For a given record to be selected, it must meet the active criteria, unless otherwise specified by the arguments in the @DBISSEL @function. The optional criteria in the @function takes precedence over the active criteria unless you start the optional criteria with a logical #AND# or #OR# operator.

DATABASE @DB @FUNCTIONS

Database @DB @functions act on an entire database file (@DBOPEN and @DBCLOSE), manipulate or display data within a database file (@DBUPD and @DBFLD), or locate specified records in a database file (@DBFIRST, @DBNEXT, and @DBPREV).

@DBAPP(ALIAS)

Adds a new blank record to the database. Records are always appended at the end of the data file. @DBAPP returns the record number of the newly appended record. If the record could not be appended, a 0 is returned.

Once a new record has been appended with @DBAPP, the fields in the record can be assigned values using @DBUPD.

@DBAPP is not available in some versions of @BASE.

@DBCLOSE(ALIAS)

Closes the database file ALIAS and returns a 1. If the database cannot be closed, @DBCLOSE returns a 0. @DBCLOSE performs the same operation as the @BASE ***F***ile ***C***lose menu command.

@DBDEL(ALIAS,RECNO,DEL_FLAG)

Sets the delete flag for a given database record. ALIAS is the alias of the database. RECNO is the record number of the record to be affected. DEL_FLAG is @TRUE (non-zero) if the record is to be marked for deletion, @FALSE (0) if the record is to remain active.

@DBDEL is not available in some versions of @BASE.

@DBFIRST(ALIAS,CRITERIA)

Returns the record number of the first record found in the database file ALIAS that meets the active and/or optional criteria.

@DBFIRST always starts searching the database from the beginning of the file. If no active or optional criteria are specified, the record number of the first record in the file is returned.

If the @BASE deleted setting is set to ignore deleted records, then @DBFIRST does not locate records that are marked for deletion. The record number of the next unmarked record that meets the active criteria after the deleted record is returned. If no record is found matching the specified criteria, an ERR value is returned.

```
@DBFLD(ALIAS,FIELDNAME,RECNO)
```

Returns the contents of the field FIELDNAME contained in the record RECNO.

Date fields return the stored serial date number, but you can set up cells to display the date in the MM/DD/YY format.

Logical fields return a 1 (True) for fields containing a T, or a 0 (False) for fields containing an F or fields that are undefined (NA).

Any field from any record can be obtained with the @DBFLD @function, regardless of the criteria and deleted settings.

```
@DBFN(ALIAS,FIELDNO)
```

Returns the field name of the field specified with the FIELDNO argument.

```
@DBLAST(ALIAS,CRITERIA)
```

Returns the record number of the last record in an open database that matches the active and/or optional criteria. @DBLAST is not available in some versions of @BASE.

```
@DBMAIN(0)
```

Returns a 1 if the main menu of @BASE is currently on the display and returns a 0 otherwise. @DBMAIN is not available in some versions of @BASE.

```
@MEMORY(0)
```

Returns the amount of available memory at the time the @function is executed. @MEMORY requires the dummy argument (0).

```
@DBNEXT(ALIAS,CRITERIA)
```

Returns the record number of the next record in the database file ALIAS matching the active and/or optional criteria.

@DBNEXT always starts searching from the current position in the database file. For instance, if you have just finished editing record number 26 and you perform the @DBNEXT @function, the record number returned is the next record after 26 matching the criteria. If no active or optional criteria are specified, the record number of the next active record in the file is returned.

@DBOPEN(DATABASE,ALIAS)

Opens the database file DATABASE with an alias of ALIAS, and returns a 1. If the file does not exist or is already open, the @function returns a 0. This function performs the same operation as the ***F***ile ***O***pen menu command.

If you do not supply an extension with the file name, .DBF is assumed. Any extension other than .DBF must be explicitly supplied. For database files that do not reside in the Lotus default directory, a path must be supplied.

@DBPOST(ALIAS)

Writes any changes that have been made to the ALIAS file to disk. Because @BASE uses buffers to do reads and writes to disk, there may be information in buffers that has not yet been written to disk. If power is lost, missing data are possible. This @function is similar to closing and reopening a database file except that it retains current criteria.

@DBPOST returns a 0 if the operation was successful; otherwise it returns an ERR.

@DBPOST is not available in some versions of @BASE.

@DBPREV(ALIAS,CRITERIA)

Returns the record number of the previous record in the database file ALIAS matching the active and/or optional criteria.

@DBPREV always starts searching from the current position in the database file.

If @BASE is set to ignore deleted records, the @DBPREV @function checks the previous record that has not been marked for deletion.

@DBUPD(ALIAS,FIELDNAME,RECNO,VALUE)

Replaces the contents of the field FIELDNAME in the record RECNO with the field value VALUE, and returns a 1. If @DBUPD is unable to replace the field, a 0 is returned.

Fields can be updated for any records in the active database, regardless of the active criteria or whether the records are marked for deletion. @DBUPD ignores any criteria or deletion settings.

A 0 is returned if you attempt to fill a field with an incorrect data type or if you supply an invalid field name or alias.

Logical fields are updated with @DBUPD by using a 1 (True) and 0 (False). For example, if you wish to change a field from F (False) to T (True), you would replace the field with a 1 (True).

With some versions of @BASE, a record is appended to the end of the database if you use a record number that is greater than the number of records in the file, or is zero.

STATISTICAL @NDX @FUNCTIONS

Statistical @NDX @functions are similar to @DB @functions. They compute values based on numeric data in an open database file. Each @NDX @function returns a numeric value.

```
@NDXAVG(DATABASE ALIAS,FIELD NAME,SEARCH VALUE,[CRITERIA])
```

Returns the average of the values in the numeric field FIELD NAME for all records with the given SEARCH VALUE that also meet the optional criteria. Records must have an exact match on the index expression of the current primary index to be included in the computation. Records marked for deletion are not included if @BASE is set to ignore deleted records.

```
@NDXCNT(DATABASE ALIAS,SEARCH VALUE,[CRITERIA])
```

Returns a count of the number of records with an index value equal to SEARCH VALUE that also meet the optional criteria. Records must have an exact match on the index expression of the current primary index to be included in the count. Records marked for deletion are not included if @BASE is set to ignore deleted records.

```
@NDXDATE(LOTUS SERIAL DATE)
```

Returns the dBASE III serial date for the specified Lotus serial date. @NDXDATE is required when you want to use a cell reference to perform a lookup on a date index.

There is a difference in the way that Lotus and dBASE III store dates. Both products store the date internally as a serial number. Lotus stores each date as a serial number representing the number of days since January 1, 1900. dBASE III stores each date as the number of days since January 1, 0001.

Normally, the difference in date values is not a problem. When importing and exporting data between a worksheet and DBF file, @BASE performs the appropriate conversion. You may, however, want to use a cell reference to perform a lookup on a date index. In that case, and only in that case, you will need to adjust for the difference in the reference dates between the two programs.

To find a match on the date, you must add a constant value of 2415019 to the Lotus serial date. The @NDXDATE @function will add the constant for you.

```
@NDXMAX(DATABASE ALIAS,FIELD NAME,SEARCH VALUE,[CRITERIA])
```

Returns the maximum value in the numeric field FIELD NAME for all records with the given SEARCH VALUE that also meet the optional criteria. Records must have an exact match on the index expression of the current primary index to be included in the computation. Records marked for deletion are not included if @BASE is set to ignore deleted records.

@NDXMIN(DATABASE ALIAS,FIELD NAME,SEARCH VALUE,[CRITERIA])

Returns the minimum value in the numeric field FIELD NAME for all records with the given SEARCH VALUE that also meet the optional criteria. Records must have an exact match on the index expression of the current primary index to be included in the computation. Records marked for deletion are not included if @BASE is set to ignore deleted records.

@NDXSUM(DATABASE ALIAS,FIELD NAME,SEARCH VALUE,[CRITERIA])

Returns the sum of the values in the numeric field FIELD NAME for all records with an index value equal to SEARCH VALUE that also meet the optional criteria. Records must have an exact match on the index expression of the current primary index to be included in the computation. Records marked for deletion are not included if @BASE is set to ignore deleted records.

DATABASE @NDX @FUNCTIONS

@NDX @functions act on index files (@NDXOPEN, @NDXCLOSE and @NDXSELECT), construct character expressions (@NDXPAD), locate specified records in an index file (@NDXFIRST, @NDXNEXT, and @NDXPREV), or open view files (@VIEWOPEN).

@NDXCLOSE(DATABASE ALIAS,INDEX ALIAS)

Closes an open index file like the *I*ndex *C*lose menu command. Returns a 1 if the index file is successfully closed; otherwise returns a 0.

@NDXFIRST(DATABASE ALIAS,SEARCH VALUE,EXACT MATCH FLAG,[CRITERIA])

Returns the record number of the first record in a data file with an index key field that matches the supplied search key value.

This @function is also known as an index seek. It uses the primary index file to look for the first record with a key value equal to SEARCH VALUE. Its function is similar to using @DBFIRST with a criteria. However, it is much faster because it uses the primary index to perform the lookup. @NDXFIRST can only be used to look for a record with the specified index key value. It cannot be used to search for a match on other field values.

If the exact match flag is @FALSE, the @function searches for the first index key greater than or equal to the specified search key value, otherwise the @function searches for an exact match to the search key value. The default setting of this flag is @TRUE.

The format of the search key value must be the same as that of the index expression.

Following a succcessful @NDXFIRST, you can use the @NDXNEXT @function to retrieve the next record that matches the search value, and @NDXPREV to retrieve the preceding matching record.

@NDXNEXT(DATABASE ALIAS,SEARCH VALUE,[CRITERIA])

Returns the record number of the next record in a data file with an index key field that matches the supplied search key value.

The @NDXNEXT command uses the primary index file to look for the next record with a key value equal to SEARCH VALUE. @NDXNEXT can only be used to look for a record with the specified index key value. It cannot be used to search for a match on other field values. @NDXNEXT always searches for an exact match on the key value.

The format of the search key value must be the same as that of the index key expression. If there are no records with the specified search value that pass the criteria, @NDXNEXT returns ERR.

@NDXOPEN(DATABASE ALIAS,INDEX FILENAME,[INDEX ALIAS])

Activates an index file for an open database file. The index file is assigned the alias specified in the last argument. The alias defaults to the file name without the extension if one is not specified. @NDXOPEN returns a 1 if @BASE is able to open the index; otherwise a 0 is returned. This @function performs the same operation as ***I***ndex ***O***pen.

@NDXPAD(DATABASE ALIAS,FIELD NAME,FIELD VALUE)

Constructs a character expression from the field value specified. The field value will be adjusted to conform to the format expected by the current primary index. The intent of @NDXPAD is to aid in the construction of index expressions involving multiple fields. The operation performed depends on the field type and current global settings as noted in Figure D-1.

@NDXPREV(DATABASE ALIAS,SEARCH VALUE,[CRITERIA])

Returns the record number of the previous record in a data file with an index key field that matches the supplied search key value.

The @NDXPREV command uses the primary index file to look for the previous record with a key value equal to SEARCH VALUE. @NDXPREV can only be used to look for a record with the specified index key value. It cannot be used to search for a match on other field values. @NDXPREV always searches for an exact match

Field Type	Value Returned
CHARACTER, with Case set to *U*se-case.	Padded on the right with spaces.
CHARACTER, with Case set to *I*gnore-case.	Padded on the right with spaces and converted to upper-case.
NUMERIC	Right-justified and padded on the left with spaces.
DATE	Character string in the form YYYYMMDD with month and day space-filled to the left.

Figure D-1

on the key value.

The format of the search key value must be the same as that of the index key expression. If there are no more records with the specified search value that pass the criteria, @NDXPREV returns ERR.

```
@NDXSELECT(DATABASE ALIAS,INDEX ALIAS)
```

Selects the primary index file for an open database file like the *I*ndex *S*elect menu command. @NDXSELECT returns a 1 if the specified index file is open and was made the primary index; otherwise a 0 is returned.

```
@VIEWOPEN(VIEWFILE NAME)
```

Opens the target data file as well as any lookup files and indexes, and activates the virtual fields (joined and computed) associated with a saved view file. @VIEWOPEN operates like the *V*iew *O*pen menu command, but returns a 1 if the view file is successfully opened; otherwise it returns a 0.

@BASE Operating Function Summary

The @BASE Option Pac supports a number of functions that can be used when creating index key expressions and computed fields. The functions are categorized as string, numeric, date, logical, and database functions.

The syntax of these functions differs slightly from the @function syntax. All functions have opening and closing parentheses after the function name. The parentheses may or may not contain data to be evaluated. Only character strings are set within quotes. Field names, numeric values, and logical values are not quoted. Either single or double quotes may be used, but they must be paired consistently.

Operating functions can only be used in index key expressions and computed fields. When constructing index expressions in a spreadsheet, the equivalent Lotus @functions must be used.

STRING FUNCTIONS

INSTR

INSTR checks to see whether a specific substring is found anywhere in the string. If the substring is found, the function returns an integer which points to the starting position of the substring within the string. If it is not found, the function returns a 0.

LEFT

LEFT extracts the leftmost portion of a character string. LEFT has two arguments: the field which contains the string and the number of characters to extract.

LEN

LEN returns the length of the character string.

LOWER

LOWER converts a string to lower-case.

LTRIM

LTRIM removes any blanks or spaces at the beginning of a character string.

RIGHT

RIGHT extracts the rightmost portion of a character string. The function has two arguments: the string and the number of characters to extract. RTRIM is usually used in conjunction with RIGHT to remove blanks at the end of the field.

RTRIM

RTRIM removes any blanks or spaces at the end of a character string.

STR

STR converts a numeric expression to a character expression. This function has three arguments: the numeric expression, the length of the converted string, and the number of decimal places. If the length is not specified, the string length defaults to 10 characters. If no decimal places are specified, the number is rounded to an integer.

SUBSTR

SUBSTR extracts a substring from a string. This function has three arguments: the string, the starting position, and the number of characters to extract. If the third argument (the number of characters to extract) is omitted, characters are extracted until the end of the expression is reached.

TRIM

TRIM operates the same as RTRIM.

UPPER

UPPER converts a string to upper-case.

VAL

VAL converts a character string composed entirely of numbers to a numeric expression.

NUMERIC FUNCTIONS

ABS

ABS returns the absolute value of a numeric expression.

INT

INT returns the integer part of a numeric expression without rounding.

ROUND

ROUND allows you to round a numeric expression by a specified number of decimal digits. This function has two arguments: the numeric expression and number of decimal places to return in the rounded number.

DATE FUNCTIONS

CTOD

CTOD converts a string to a date variable. The string may be in any date format supported by LOTUS.

DAY

DAY returns the day portion of a date expression. The day is expressed as an integer.

DTOC

DTOC converts date format to a string in the Lotus default date format.

MONTH

MONTH returns the month of a date expression. The month is expressed as an integer in the range 1 through 12.

YEAR

YEAR returns the year portion of a date expression. The year is expressed as a four-digit number.

TODAY()

TODAY() returns the Lotus serial date number corresponding to the current system date.

LOGICAL FUNCTIONS

DEL()

DEL() is a logical function that returns "T" if a record is marked for deletion or "F" if the record is not marked for deletion. DEL() can be used in computed field expressions but cannot be used in index key expressions.

IF

IF returns one of two values depending on the value of a logical expression. The function has three arguments: the logical expression, the value to be returned if the expression is true, and the value to be returned if the expression is false.

IIF

IIF operates the same as IF.

DATABASE FUNCTIONS

RECNO()

RECNO() returns the current record number as its value. A record number is the physical position of a record in the selected database file. RECNO() cannot be used in index key expressions.

Index

Thank you for purchasing this book. The author would like to offer you the opportunity to obtain the applications and examples in this book on diskette. To save keyboard input time, all of the applications and examples in this book are available on disk. Each application has the original template(s), named ranges, program(s), etc., described in the book. The applications will work with Lotus 1-2-3®, versions 2.0/2.01/2.2 and Lotus Symphony®, versions 1.2/2.0/2.2. Also available are the add-in application programs described in the book. To send for the ***Practical Laboratory Information Management For Scientists & Engineers*** disks and/or the application programs, just photocopy this page, fill out the information completely, and return to the address listed below. Please make checks payable in U.S. dollars to Louis M. Mezei. Thank you.

QUANTITY

Practical Information Management For Scientists and Engineers

_____ $48.00 plus $2.00 shipping and handling.

_____ Companion disks: $38.00.

Laboratory Lotus®: A Complete Guide To Instrument Interfacing

_____ $48.00 plus $2.00 shipping and handling.

_____ Companion disks: $18.00.

Practical Spreadsheet Statistics and Curve Fitting For Scientists and Engineers

_____ $48.00 plus $2.00 shipping and handling.

_____ Companion disks: $18.00.

_____ *Personics @BASE*: $195.00 plus $2.00 shipping and handling.

_____ *Personics @BASE Option Pac*: $99.95 plus $2.00 shipping and handling.

_____ *Lehrman Associates: EXTRA K*: $79.95 plus $2.00 shipping and handling.

_____ *Intex Solutions: BEYOND 640*: $95.00 plus $2.00 shipping and handling.

Please specify the Release of Lotus software and disk size that you use:

_____ *Lotus 1-2-3 Release*

_____ *Lotus Symphony Release*

_____ *Disk Size (3.5" or 5.25")*

Date Purchased: _______________

Name: ____

Title: ____

School/Company: ______________________________________

Department: ____

Street Address: _______________________________________

City: ___________________ State: _________ Zip: ____________

Telephone: (   ) _______________________________________

Louis M. Mezei
40815 Ondina Ct.
Fremont, CA 94539